# Basic Biostatistics

## *Concepts for the Health Sciences*

*The Almost No Math Stats Book*



**Frank D. Bell**
*Indiana State University*

WCB Wm. C. Brown Publishers
Dubuque, Iowa • Melbourne, Australia • Oxford, England

**Book Team**

Editor *Colin H. Wheatley*
Developmental Editor *Kristine Noel*
Production Editor *Karen L. Nickolas*
Art Editor *Joseph P. O'Connell*
Publishing Services Coordinator/Design *Barbara J. Hodgson*

**Wm. C. Brown Publishers**
A Division of Wm. C. Brown Communications, Inc.

Vice President and General Manager *Beverly Kolz*
Vice President, Publisher *Kevin Kane*
Vice President, Director of Sales and Marketing *Virginia S. Moffat*
Vice President, Director of Production *Colleen A. Yonda*
National Sales Manager *Douglas J. DiNardo*
Marketing Manager *Craig S. Marty*
Advertising Manager *Janelle Keeffer*
Production Editorial Manager *Renée Menne*
Publishing Services Manager *Karen J. Slaght*
Permissions/Records Manager *Connie Allendorf*

**Wm. C. Brown Communications, Inc.**

President and Chief Executive Officer *G. Franklin Lewis*
Corporate Senior Vice President, President of WCB Manufacturing *Roger Meyer*
Corporate Senior Vice President and Chief Financial Officer *Robert Chesterman*

Cover and interior design by Anderson Creative Services



A Times Mirror Company

Library of Congress Catalog Card Number: 93–73781

ISBN 0–697–14979–X



Printed in the United States of America by Wm. C. Brown Communications, Inc., 2460 Kerper Boulevard, Dubuque, IA 52001

10 9 8 7 6 5 4 3 2 1

*To Susan, Josh, Mom, Bob, and the memory of my father.*

# CONTENTS

## 3 Samples and Populations 19



## 4 Searching for Answers: Experimentation and Observation 33



## 5 Describing Data 43

# PREFACE

The purpose of this book is twofold: 1) to enable students of the health sciences to apply introductory descriptive and inferential statistical procedures to problems in their field, and 2) to enhance their understanding and critical review of related professional literature. The emphasis throughout is to provide explanations for the basic concepts upon which nearly all of the most common statistical techniques are founded. Whenever possible, the mathematical derivations of statistical formulas and the use of complicated computational formulas have been avoided. Instead, the focus stays on the broad concepts.

This book was written for beginning students in biostatistics, for students who are non-mathematics majors, and for those students who may profess math anxiety. The language used throughout the text has been developed over 17 years of teaching introductory statistical concepts to those students who represent health-related disciplines such as nursing, medicine, social work, health education, environmental health, health and safety management, athletic training, physical education, and physical therapy. The tables, figures, examples, and the application and comprehension problems have been refined by classroom use and professional review to make them appropriate for the typical student of introductory statistics. Some calculations that involve elementary algebra are demonstrated in later chapters of the book. In every case, these calculations are clearly presented and explained in a step-by-step fashion that promotes student confidence and understanding.

The chapters of the book have been organized to serve equally well both as a text for self-directed study and for a traditional lecture course. This flexibility stems from the orderly development of each chapter's content. For example, the chapter reviews and application problems (with answers) cultivate student learning and instill confidence using only minimal mathematical calculations. The uncomplicated tables and figures support content development and serve to reinforce learning.

Probability tables have been dramatically modified to be more user-friendly. The examples used are straightforward, timely, and represent a variety of health-related disciplines. In general, the book is organized as a concise, clear, understandable, and nonthreatening introduction to statistics.

I was very fortunate to have a knowledgeable and hardworking panel of reviewers, whose forthright criticisms and helpful suggestions added immeasurably to the quality of this first edition. The review panel included:

Michael S. Bisesi
Medical College of Ohio

Gerald C. Hyner
Purdue University

Bob Ford
Ball State University

Timothy J. Smith
University of the Pacific

CHAPTER

**1**

# Introduction: Why Study Statistics?

*"It is better to know some of the questions than all of the answers."*

**JAMES THURBER**

Why do students study statistics? The resounding response from undergraduate students in most introductory classes is: "A course in statistics is required for my major." Most graduate students respond similarly but are quick to add, "I will need a statistics course to pass my competency exams," or "I need a statistics course to help me with my graduate research experience." A better question might be, "Why is the study of statistics required as part of the academic preparation for students of many disciplines, especially those students preparing for health-related careers?" A statistician would probably answer in this way: "The study of statistics enables students to collect, organize, analyze, interpret and present data in order to make clear, concise, and objective decisions." Yet often, beginning students have trouble understanding what it is they need to decide.

Ultimately, decisions are made to answer questions, implied or stated, concerning the information available. For example, the community health educator might ask, "Which community health education programs can be developed to meet the needs of communities like mine?" The use of statistical methods involved in the collection, organization, analysis, and interpretation of available information can help the community health educator decide not only which programs are needed, but which programs are needed *most* and which of these have the greatest chance for success. The health educator, in this example, approaches the question deductively (reasoning from the general to the specific) by considering the general program needs of many communities similar to his or her specific community before developing specific programs. Or, the health educator approaches the problem inductively (reasoning from the specific to the general), examining the needs of the specific community in question and developing a programmatic model for all similar communities. Thus, understanding statistical methods benefits both the *deductive* and *inductive* approaches to decision making.

## TYPES OF STATISTICAL METHODS

Statistics is defined in two ways. The first definition tells us how we think of statistics. That is, statistics are numbers resulting from a collection of data. This definition is important, but the second definition will be used more often in this book. That is, **statistics** is the method of *describing and interpreting data.* Data are important regardless of which definition of statistics we use. For now, we will define **data** as the set of measurable, available information. These sets of measurable, available information (data) are subject to variability and are, therefore, classified as **variables.** Examples of variables include age, gender, systolic blood pressure, and patient satisfaction. We will continue our discussion of data and variables in Chapter 2.

The second definition of statistics indicates that statistical methods are divided into two separate disciplines: descriptive statistical methods and inferential statistical methods. **Descriptive statistical methods** include the summarizing, condensing, and organizing of sets of data into tables, figures, and numbers. These descriptive methods can serve to manage sets of data, reducing them into more understandable ''pictures'' and terms. **Inferential statistical methods** are the methods used for answering questions about the general nature of all things using the measurable, available information (data) collected from a few things. Inferential methods predict or estimate characteristics of the *whole* based on characteristics of a *part* or subset of the whole.

Statistical methods are tools of the research process; as such, these methods can help us make decisions both deductively and inductively.

## STATISTICAL METHODS AND DEDUCTION

The study of statistical methods can be of value in the application of general principles to a specific situation. Familiarity with statistical methods and knowledge obtained from academic preparation, from experience, and from review of the literature can help you reach the best answers possible for your particular client, patient, worker, or situation.

This deductive approach to decision making is utilized in many professions, especially health-related professions. For example, a specific patient, client, or worker may ask, ''My total cholesterol level was reported to be 250 mg/dL. What are my chances of heart disease?'' Or, the boss may ask, ''Have our safety programs been effective when compared to the industry in general?''

In the first case, the patient or client asked for *general principles* of cardiovascular disease to be applied to his or her *specific* case. To answer the question, general information including age, sex, and family history may be considered. The collection, organization, analysis, interpretation, and presentation of information from appropriate literature sources is required before a ''best,'' if not ''correct,'' answer can be given.

In the second example, the boss has asked how, in terms of the safety records of the industry in general, the safety records of the specific company within that industry compare? **Quantitative** and **qualitative data** (available information) including, perhaps, indicators of accidental injuries, cost, lost time, production, and worker satisfaction may need to be considered to answer this question. What is learned about the safety records of the industry in general may then be compared to the safety records of the specific company. This requires the collection, organization, analysis, and interpretation of information available from appropriate sources before these data can be presented concisely to the boss.

These two examples serve to illustrate how the understanding of statistical methods can enhance the deductive approach to decision making. The application of statistical methods to information gained concerning the general nature of things is an important component of the deductive process. One must always keep in mind, however, that the general nature of things is susceptible to change. If we allow ourselves to habitually interpret the general nature of things in the same way, we may be prone to faulty reasoning and risk committing the same mistakes over and over again. For example, the best answers to the questions in the previous examples may be different today than they were 10 years ago; and the best answers given to these same questions next year might be different than the best answers we can give for these same questions today. It is not that the questions change; it is that the information available concerning the general nature of things may change. Ten years ago the perceived role of blood levels of cholesterol in the determination of cardiovascular disease was different than today. Next year, the effectiveness of the industry's safety programs may be changed as a result of some other factor. The application of statistical methods to the general nature of things helps the professional keep track of changes that influence his/her profession.

## STATISTICAL METHODS AND INDUCTION

In our work, we often encounter situations in which we need to make decisions concerning questions that arise from new information or from information that does not have a clear link to our previous experience. In these situations we must conduct experiments or other types of studies to help answer these questions.

For example, to answer a new question, we may decide to study a small group of subjects and then, based on what we observe, attempt to make inferences concerning what will happen to all such subjects, in general. This method of answering questions is known as **inductive reasoning** and typically involves observing a few specific subjects in order to better understand what might happen to a general group of like subjects—reasoning from the specific to the general.

Inductive reasoning is popularly used in health-related research to provide an ongoing redefinition of the general nature of things. For example, the medical researcher may be interested in understanding what side effects result from the use of a new drug in the treatment of a particular condition or disease. To gain this understanding, the researcher may decide to prescribe the drug to a small group of patients exhibiting symptoms of the condition or disease and observe the results. After observing the results of the experiment, the researcher may attempt to define what side effects to expect if the drug was used to treat all such patients exhibiting such symptoms.

On the surface, it would appear to be easy to reach conclusions regarding the potential side effects of the new drug on patients exhibiting symptoms of the disease or condition. Yet, as we consider the experiment more deeply, we are faced with more questions. Do side effects occur equally in both male and female patients? Are younger patients more likely to exhibit side effects? Are the side effects uniform from patient to patient? Does the seriousness of the disease or condition influence the expression of side effects? One way researchers have attempted to objectively organize their inductive research efforts is by using the scientific method, which has five basic parts:

1. *Questioning*—Stating the question to be researched.
2. *Observing*—Collecting, organizing, and interpreting available information (data) related to the question.
3. *Formulating a Hypothesis*—Creating a plausible answer to the research question suggested by the available data.
4. *Designing the Study or Experiment*—Creating a method that can be used to test the hypothesis, determining what data will be considered and how.
5. *Conducting the Experiment or Study*—Observing the outcomes of the experiment and organizing and interpreting the results in such a way that the original hypothesis is either accepted or rejected.

Once we have conducted the experiment, we have results which can be used to reject or accept the original hypothesis. If our results seem to contradict the original hypothesis, we begin the process again and generate a new hypothesis to be tested. If, on the other hand, our results seem to affirm the original hypothesis, then we can apply what we have learned to the general nature of things. We must, however, guard against accepting the affirmed hypothesis as the absolute truth and realize that only in light weaknesses of the experimental design and the current available knowledge does the hypothesis appear to be true.

Knowledge of statistical methods can serve as an invaluable resource at each step of the inductive process and can help you avoid the pitfalls associated with data collection, analysis, presentation, and interpretation. Such knowledge can aid in the development of study designs for testing hypotheses, help you make better decisions, and ultimately save you time and heartache.

## Deductive and Inductive Inference

As we have pointed out, decisions can be made deductively or inductively, and an understanding of statistical methods can enhance both approaches to decision making. Both forms of reasoning are used to make inferences.

With **deduction,** the inference is made from what can be learned about the general nature of things to the nature of a specific thing in question. For example, the environmental health specialist might be interested in the life cycle of a particular local insect known to carry a pathogenic agent. By examining the available information related to all insects of this particular type in similar locations, the health specialist will be able to make a very accurate inference concerning the life cycle of this particular insect. If reasonable methods are used, deductive inference can be a highly accurate, low-risk proposition.

With **induction,** the inference is made from what can be learned about a small number of specific things to the general condition of all such things. For example, imagine that the environmental health officer is interested in the effectiveness of a newly developed pesticide when the pesticide is used to kill a particular type of insect. To test the effectiveness of the pesticide, the health specialist would probably not be inclined to attempt to kill all such insects in all locations worldwide. Instead he or she might decide to collect a few of these insects, put them in a cage, spray them with the pesticide, and observe what happens. If each of the insects are killed, should the environmental health specialist claim that the pesticide is absolutely effective to kill all such insects worldwide? In trying to generalize the results of his experiment you might ask these questions: If an additional batch of 30 such insects were sprayed with the pesticide would the results be the same? Would the results be the same if the insects came in contact with the pesticide in some way other than being sprayed? Would the results be the same at different ambient temperatures or levels of relative humidity? As you can see, it is often very difficult to prove anything with absolute confidence using induction. Making inferences from the specific to the general is often a high-risk proposition.

When you attempt to interpret the results of inductive studies, conducted by yourself or others, you must ask if the specific group studied fairly represents the larger group to which inference is made. If not, the results should be considered with caution. A knowledge of statistical methods can help you protect yourself from misinformation and/or misinterpretation.

## SUMMARY

A knowledge of statistical methods provides us with a mechanism to answer questions that arise in our work and private lives. Moreover, this knowledge can help us avoid the misinterpretation of available information. It helps us become more critical of our own work and the work of others. Ultimately, the application of statistical principles and techniques to questions that we face on a daily basis not only helps us come up with reasonable answers but, more importantly, helps us recognize additional questions in need of answers.

## FOR REVIEW

1. Statistics deals with the methods of collecting, organizing, analyzing, interpreting and presenting available information that can be expressed numerically.
2. A statistic is (statistics are) usually a number (or numbers) used as an estimate of an unknown value. Later, we will explain that a statistic is (statistics are) a number (or numbers) describing a sample.
3. Data includes the set of measurable, available information that will be statistically considered. These data, available information, usually possess some degree of variability. Statistical methods attempt to summarize data trends and draw reasonable conclusions in spite of the uncertainty caused by the variability of the data. ''Datum'' is the singular, while ''data'' is the plural form of the noun. Data are often subdivided as quantitative or qualitative.
4. Quantitative data, such as height and weight, are those data that can be easily assigned a number.
5. Qualitative data, such as happiness and grief, are those data that cannot be assigned (reasonably) a number.
6. Deductive reasoning is reasoning from the general nature of things to the specific thing in question.
7. Inductive reasoning is reasoning about the general nature of things from conclusions based on specific observations.
8. The scientific method serves as a framework to objectively test new ideas in light of current knowledge.
9. The hypothesis is an educated guess based on available information concerning the outcomes of an experiment.
10. Inference involves making decisions or arriving at conclusions by reasoning either:
    a. Deductively, from the evidence of observed numerical data concerning the general nature of things to the specific; or,
    b. Inductively, from the evidence of observed numerical data concerning the specific to the general nature of things.
11. A variable is a classification or characteristic that can have more than one value. For example, gender is a variable that can have the value of 1 = male or 2 = female.

CHAPTER

**2**

# Data, Variables, and Cases

*"You can't measure time in days the way you can money in dollars because every day is different."*

**JORGE LUIS BORGES**

In Chapter 1 we discussed statistical methods and emphasized the role of statistics in the collection, organization, analysis, interpretation, and presentation of data. We also defined a statistic as numbers resulting from a collection of data; and data as measurable, available information, either quantitative or qualitative in nature. At this time, it is important to understand that available information (data) that can be assigned a number (quantified) in some way may be used as a statistic. The available information that may not be quantified is referred to as qualitative in nature and often does not lend itself easily to the methods of statistical analysis. Let's begin this chapter with a review of quantitative and qualitative data. Then we will consider scales of measurement, variables, and cases.

## QUANTITATIVE DATA

All of us are familiar with a variety of quantitative measures. We have included a cup of milk in our favorite recipe, measured our height with a tape measure, our weight on a bathroom scale, and noted the miles our automobile travels per gallon of fuel. In each of these cases we are reasonably comfortable converting the available data (volume of milk, height, weight, fuel efficiency) to some quantity (cups, inches, pounds, miles per gallon). These are all examples of data that can be expressed quantitatively (as numbers).

## QUALITATIVE DATA

We are also aware that some available information is difficult, at best, to quantify. For example, your level of frustration on the highway at rush hour, the quality of the instruction you received in a particular course, or the depth of sorrow felt after the loss of a loved one are all qualities which may be difficult (some would say impossible) to reduce to numerical values. Qualities such as these are often researched using verbal or written descriptions without the assignment of numbers (quantification) and are not, therefore, analyzed statistically. Qualitative researchers contend that any attempt to quantify information that is typically qualitative in nature creates an illusion of objectivity and serves to oversimplify, distort, or otherwise falsify the phenomenon under study. On the other hand, quantitative researchers are often willing to assign numerical values to qualitative data in an attempt to analyze such responses statistically.

## SCALES OF MEASUREMENT

Measurement involves the assignment of numbers to observations made from available information. Whether we step on the bathroom scale and record our weight, or record the number of times a grief victim mentions the word "sad," or note the order of finish of a race, we are assigning numbers to observations. The way in which we assign these numbers enables us to recognize which scale of measurement is being used.

Assigning numbers to observations (measurement) involves some method or rule of assignment. The method or rule chosen for the assignment of numbers to observations is the best indicator of how the measurement can be used.

In 1946, S. S. Stevens described a measurement classification system. The use of the Stevens System has become a tradition among statisticians. This system has proved to be of tremendous value in the determination of statistical methods to be used for a particular set of data in a particular situation. Four scales, arranged hierarchically by degree of complexity, are most often distinguished: nominal, ordinal, interval, and ratio.

### The Nominal Scale

**Nominal scaling,** or simply using numbers to label categories, is the least complicated order of measurement. The categories are assumed to be independent of each other or mutually exclusive. Some examples of categories that could be assigned numerical values are: dogs or cats; boys or girls; black, white or other; Democrats or Republicans; smokers or nonsmokers; oak or maple; Group 1 or Group 2; and I.D. numbers (including Social Security, telephone, zip code, student, and the like). Nominal scales can also be used to categorize traits or qualities based on some protocol. For example, managers could be classified as exhibiting passive or aggressive management styles after responding to a management-style survey instrument. Or, a patient or client could be classified as either a Type A or Type B personality after an interview.

The important thing to remember is that the nominal scale represents an "*either-or*" proposition. The placement of an observation in one category precludes placement in another. For example, consider the following item from a student information questionnaire:

---

Gender: If male, mark "1"
If female, mark "2"

---

To respond, students must mark either 1 or 2 on their scan sheet. By doing so, the students have classified themselves as *either* 1, *or* 2—(male or female). It becomes possible, then, to group all the males together and, similarly, group all the females together. The variable, gender, was used to categorize the students as either male or female.

For the nominal scale, there need not be a relationship between the categories of the variable being measured. Within any category of a variable (for example, male is one category of the gender variable), the members of that category are assumed to be equal in terms of the trait being measured (males in one group, females in the other).

The names, numbers, or other symbols used to classify the variable may be changed and the resulting information provided by the scale remain the same. For example, the item from the student information questionnaire could have been changed to read:

---

Gender: If male, mark "A"
If female, mark "B"

---

We can still divide all the responses to the questionnaire into two groups ("A" and "B"), males and females.

The nominal scale has been called the counting or grouping scale. We can group or count boys and girls or dogs and cats to categorize data of interest.

## The Ordinal Scale

The second measurement scale, the ordinal scale, is similar to the nominal scale in that either-or categories are established. However, the **ordinal scale** differs from the nominal scale in that the categories specified are ranked in some order (often greater than, less than, or equal to).

Let us imagine we have used two categories to classify smoking behavior. Using the nominal scale, we can classify smoking behavior as either smokers or nonsmokers. This would allow us to group smokers and nonsmokers for purposes of comparison. Suppose also, additional information was required. What if we were interested in how much smoking took place among the members of the smoking category. We could then use the ordinal scale to arbitrarily group the smokers as frequent or occasional smokers. To do this, we could define occasional smokers as those who smoke no more than one cigarette a day. Frequent smokers could be defined as those who smoke more than one cigarette a day. For example:

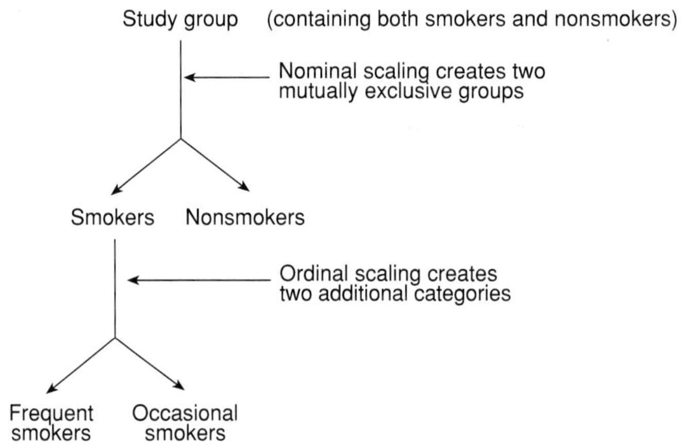


Ordinal scaling must satisfy two conditions: the mutually exclusive (same-as-or-different-than) condition and the greater-or-less-than condition. Observations that can be placed in rank order, such as race finish order (1st place, 2nd place, 3rd place, etc.), socioeconomic status rankings (upper, middle, lower), and age group ranks (young, middle-aged, old), are examples that satisfy the conditions for ordinal scaling.

Remember that ordinal scales provide information concerning quantity—faster than or slower than, more than or less than, and younger than or older than—but they do not tell you *how much* faster than, *how much* more than, or *how much* younger than. It is true that, with ordinal scaling, we assume the existence of a continuum underlying the numbers. However, it is impossible to know exactly, for example, how much faster the 1st place winner was than the 2nd place winner. With ordinal scaling, we only know who was 1st and who was 2nd; we cannot determine if it was a photo finish or if the 1st place winner had showered, changed clothes, and had dinner before the 2nd place winner finished. In ordinal scaling, as in this example, there is no known interval between 1st and 2nd, just a known order. One common mistake we are often tempted to make when we interpret ordinal scales is to assume that the differences between the ranks are equal. Consider the following example:

Suppose you are interested in purchasing the most reliable, state-of-the-art, sound level meter on the market. After a visit to the library where you have collected available information concerning sound level meter cost and reliability, you are faced with the following information (assume that each of these models meets your needs otherwise):

| **Model** | **Reliability Ranking** | **Cost** |
|---|---|---|
| #1 | First | $600.00 |
| #2 | Second | 400.00 |
| #3 | Third | 200.00 |

If you had $600 that you were willing to spend on a sound level meter, and if you were interested in buying the sound level meter that was ranked first in terms of reliability, which sound level meter would you select? Your choice can be made easily.

But, if you were able to spend $600, and you were interested in buying the most reliable sound level meter *for the money,* which would you select?

If just before you made the purchase you were given additional information, how might this information influence your purchase?

| **Model** | **Reliability Ranking** | **Months of Service Without Repair** | **Cost** |
|---|---|---|---|
| #1 | First | 48 months ± 1 month | $600.00 |
| #2 | Second | 46 months ± 2 months | 400.00 |
| #3 | Third | 14 months ± 6 months | 200.00 |

Before you were provided the additional information concerning months of service without repair, you were faced with the prospect of making a decision based simply on the reliability rankings of the models in question. You might have been tempted to consider the costs and select model #1 as the most reliable *and* as the most reliable for the money. You may have assumed, after considering the rankings and the costs, that the difference in reliability between model #1 and model #2 was equal to the difference in reliability between model #2 and model #3. This assumption may not be warranted when considering the reliability rankings and the costs alone. To illustrate this, reconsider the second set of information.

The second set of information provided us with the number of months each model functioned without repair. Model #1 functioned for 47 to 49 months (48 months, plus or minus one month) without repair and cost $600. Model #2 functioned for 44 to 48 months without repair and cost $400. Model #3 functioned for 8 to 20 months without repair and cost $200. If we want the most reliable sound level meter, regardless of cost, model #1 is the one to purchase. However, if we want the most reliable machine for the money, model #2 is the one to buy.

When we considered the first set of data presented, we were only able to estimate reliability of the models by considering the models' reliability rank. That is, model #1 ranked first, model #2 ranked second, and model #3 ranked third. Ordinal scaling contributes information concerning same-as-or-different-than and greater-than-or-less-than, but contributes nothing to our knowledge concerning the degree of difference between the ranks (keep in mind, from the smoking behavior example, the frequent smoker group, by definition, could have included smokers who smoked any number of cigarettes more than one).

The second set of data included information concerning the *degree of difference* in reliability (months of service without repair) in addition to the reliability ranks. Only after these degree-of-difference-data were considered could we conclude that model #2 was probably the best choice for the money. This degree-of-difference-data presented in the second data set allowed us to understand the intervals between ranks. In this case, the information concerning the months of service without repair represents an example of interval scaling, which is our next topic.

## The Interval Scale

The interval scale is similar to the ordinal scale in that either-or and greater-than-or-less-than categories are established. In addition, as we saw with the previous example, the interval scale is also concerned with the degree (interval) of difference. In that example, the interval nature of the data, months in service without repair, allowed us to learn how much more reliable (in months) model #1 was when compared with models #2 and #3. As a result, we were able to better determine which model was the best choice for the money.

In general, scales that are involved in measurement are often interpreted to be of the interval or the ratio scale. In the **interval scale,** numbers are assigned along an assumed continuum in such a way that the points along the continuum are of the same magnitude. Other examples of scales that provide interval data are temperature, height, weight, units of area, and volume.

The use of some scales of measurement involves beginning at a nonarbitrary, zero starting point. For example, if we plan to drive 500 miles on the first day of a vacation, we know after we have driven 250 miles that we have completed 50% of the driving for that day. Or, if we are given a written exam with 100 questions and correctly answer 80 questions we know that we have answered 80% of the questions correctly. In both of these cases we have begun at a nonarbitrary, zero starting point and our progress can be reported as some fraction or percentage of the whole. Both of these examples illustrate ratio scaling.

### The Ratio Scale

The ratio scale is similar to the interval scale but, as we have noted in Figure 2.1, the **ratio scale** not only has equal interval distances between the points on the scale, but also has a nonarbitrary, zero starting point. Altitude (distance above sea level) is a variable that can be interpreted using the ratio scale. If we know two peaks of a particular mountain range have altitudes of 4,000 feet and 8,000 feet, respectively, then we are able to state that the second peak is 100% higher than the first. As in this case, information reported as percentages or fractions use the ratio scale of measurement.

The distinction between interval and ratio scales is not of great importance from a statistical perspective, since the same statistical methods can be applied to both types of variables.

## SCALING IN REVIEW

As we continue our investigation of statistical methods in this text, you will soon realize the statistical methods we discuss have often been designed for the analysis of distinct scales of data: nominal, ordinal, interval, and/or ratio. Many of the conventional statistical methods have evolved for use with the interval and ratio scales. We will focus primarily on the traditional methods for analysis and interpretation of data that are at least interval in nature. Additionally, we will devote attention to less traditional methods for the analysis of nominal and ordinal (categorical and ranked) data. Therefore, you must try to keep all the scales of measurement in mind as you continue your study. Remember, knowledge and comprehension of the levels of measurement used with a particular set of data will assist you in determining which statistical methods have been traditionally used in the analysis of those data. Similarly, recognition of the levels of measurement used with a particular set of data presented in the literature of your profession can serve as an important first step in the process of critical evaluation.

The measurement scales discussed have been presented in order from the least informative to the most informative. Each of the scales categorize according to same-as or different-than characteristics. The ordinal scale categorizes data according to greater-than or less-than as well as same-as or different-than characteristics. Only in the interval and the ratio scales is the unit or degree of difference known. Figure 2.1 can be used to help you differentiate between scales used in the measurement of data.

How do we use Figure 2.1 to help avoid mistaking which scale is represented? Suppose a medical researcher wishes to categorize patients according to their blood pressures. The researcher determines, for the purposes of his study, that patients need to be classified as having either high blood pressure (hypertensive), normal blood pressure (normotensive), or low blood pressure (hypotensive). Assume that the researcher has reviewed the appropriate literature and has established traditional gender- and age-specific ranges

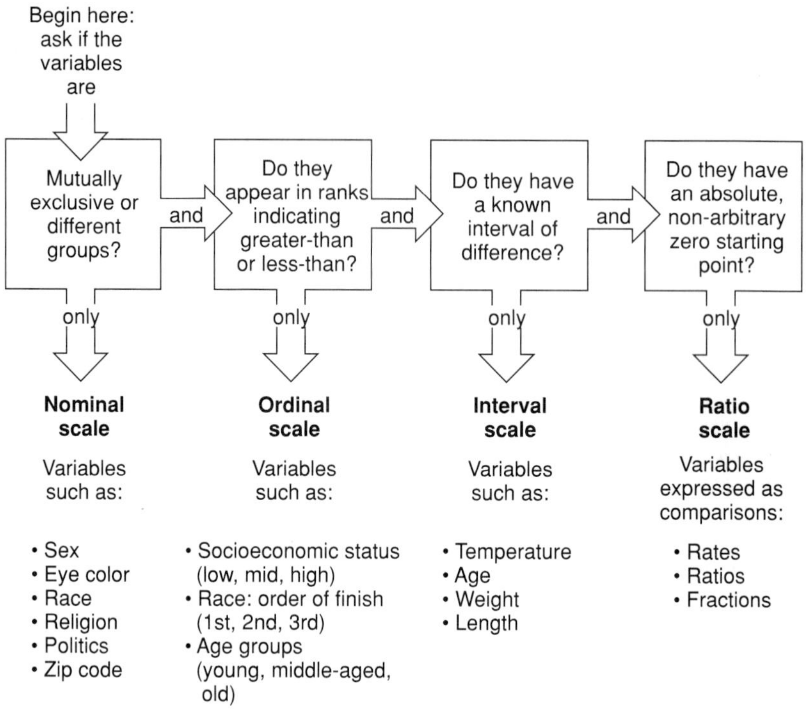


**FIGURE 2.1**
A flowchart to help identify levels of management.

that indicate the upper limits of hypotension and the lower limits of hypertension. He then classifies 100 patients as hypotensive, normotensive, or hypertensive. What scale of measurement is involved in this example?

Upon reviewing the scaling diagram we find the first question we need to ask is, "Are the variable categories mutually exclusive?" The answer to this question is "Yes" because a patient is classified as either hypotensive, normotensive, or hypertensive. At this point we know the nominal scale has been used. The next question is, "Are the categories ranked somehow indicating greater-than-or-less-than categorization?" The answer to this question is "Yes" because those patients grouped as hypotensives have lower blood pressures when compared to normotensives, and normotensives have lower blood pressures than the hypertensives. At this point, we know the variable, blood pressure status, has been ordinally scaled. The next question is, "Does the variable demonstrate a known interval or degree of difference?" The answer to this question is "No" because those patients who are hypotensive do not demonstrate a clear lower pressure limit, just as the hypertensives do not clearly demonstrate a known upper pressure limit. Additionally, there is no indication that the interval that includes the normotensives is equal to either of the other intervals. Therefore, we know that these variables are not scaled by interval or ratio scaling.

By taking the time to determine how variables have been scaled you can, with practice, overcome the tendency to attempt to ask more of the data than the data are able to tell. It is not correct to utilize nominally scaled variables and then make greater-than-or-less-than (ordinal scale) conclusions. Similarly, it is not correct to utilize ordinally scaled variables and then make how-much-greater-than or how-much-less-than (interval or ratio scale) conclusions (e.g., an increase in your socioeconomic status from the middle level to the high level should never be reported as a 33% gain).

To build a firm foundation for further study, you must realize the importance of *correctly identifying the measurement scale represented by the data in question* and understand that *particular limitations of interpretation may exist regarding the level of measurement scale used.*

## VARIABLES

As you have probably noticed, we have already used the term "variable" in the preceding discussion. In each instance, **variable** was used to indicate something that can be measured and can vary. Variables also serve to organize data. For example, available information (data) indicating whether a person is a smoker or nonsmoker can be organized under the variable "smoking behavior." Later, when it becomes more important, we will differentiate between independent variables (factors) and dependent variables (responses).

## CASES

Although we have not previously discussed cases specifically, we have alluded to examples where the cases were patients or experimental units (such as laboratory animals or equipment). In general, the **case** supplies the data that we organize as variables. It is, then, the unit of study which is the source of the variables measured. Many users of statistics find that it is beneficial to organize the relationship between the cases, data, and variables in the following fashion:

**← A Horizontal Variable Display →**

| **Case name** | **I.D.#** | **Gender** | **SES** | **Age** | **Exam %** |
|---|---|---|---|---|---|
| Bill | 01 | 01 | 1 | 20 | 92 |
| Mary | 02 | 02 | 2 | 22 | 80 |
| Ed | 03 | 01 | 3 | 21 | 77 |
| Sue | 04 | 02 | 1 | 19 | 82 |
| Joe | 05 | 01 | 2 | 20 | 68 |
| Betty | 06 | 02 | 3 | 19 | 73 |
| Allen | 07 | 01 | 1 | 19 | 94 |

In this example, six variables were measured on each individual and are displayed horizontally. The variables were: case name, identification number, gender, socioeconomic status, age, and examination percentage. The seven cases were listed vertically and included: Bill, Mary, Ed, Sue, Joe, Betty, and Allen. For example, the data indicates that Joe is 20 years old and answered 68% correct on the exam. Notice, as well, that three of the variables (case name, I.D. number, and gender) are nominally scaled, one variable (socioeconomic status) was ordinally scaled, one variable (age) was intervally scaled, and one variable (exam %) was ratio scaled.

This method of organizing data is not only useful to help gain a better understanding of the relationship between the data, variables, and cases involved in subsequent chapters, but is also very useful in the preparation of data files for the eventual analysis of the data.

## FOR REVIEW

1. Quantitative data is available information that can be expressed numerically.
2. Qualitative data is available information that may be categorized but is difficult, if not impossible, to represent numerically.
3. The nominal scale is the measurement scale that uses numbers to categorize data into mutually exclusive groups.
4. The ordinal scale is the measurement scale that orders data into greater-than or less-than ranks but does not indicate how much greater or less.
5. The interval scale is the measurement scale in which the degree of difference between successive points on the scale is assumed to be equal. Interval data indicate how much greater or how much less.
6. The ratio scale is a special interval scale for which the determination of a nonarbitrary, zero starting point is possible. Ratio comparisons can indicate that one measure is 50% less than another.
7. A variable is something that can be measured and can vary.
8. A case supplies the data that we may organize as variables.
9. Categorical data are those data that can be only grouped into mutually exclusive groups (e.g., nominal data).
10. Ranked data are those data that can be categorized (either, or) and ranked as greater-than or less-than groups (e.g., ordinal data).
11. Measurement data are those data that can be categorized into groups having a known interval of difference (e.g., interval and ratio data).

### Chapter Knowledge and Comprehension Application

For the following, indicate which scale of measurement (nominal, ordinal, interval, or ratio) is being used:

1. In 1993, in a county in Wisconsin, 10 cases of Lyme disease were reported for every 100,000 residents. In 1994, in the same county, 20 cases of Lyme disease were reported for every 100,000 residents. This indicates twice as many cases of Lyme disease than were reported the year before in that county.

2. Sixteen outpatients were treated.
3. Of 50,000 applicants for concealed-weapon permits in a certain state, 250 were turned down.
4. The phone company announced that the 317 area code serves three million customers.
5. Of the 35 most recent heart attack victims treated at a local hospital's emergency room, 5 had low total blood cholesterol levels, 10 had normal levels, and 20 had high levels of blood cholesterol.
6. The average blood pressures for these same heart attack victims was 140/94 mm.Hg.

**Answers**

1. The counts are nominal, the comparison assumes ratio.
2. Nominal scale
3. Nominal scale
4. Nominal scale
5. Ordinal scale
6. Interval scale

For the following, criticize the conclusions drawn from the examples.

1. A graduate student conducted a study to, in part, describe the available information concerning the number of children and the gender of those children in the families of a particular community. After consulting census and birth records, the student concluded that the typical family in the community had an average of 2.3 children with an average gender of 1.6.
2. A sports information director was asked to provide information concerning the effect of a recently initiated academic program for athletes. The program involved such things as special academic counseling, tutoring, and required study time. After reviewing the available information, the sports information director reported the combined grade point average (GPA) of the athletes had increased from 2.0 before the program started to nearly 3.0 after the program and concluded that this was almost a 50% increase in knowledge.

**Answers**

1. The information available concerning the number of children per family and the gender of those children was presented as nominal data (either a child or not and either male or female). The reason for the confusion was that the student used these mutually exclusive categories as if there was some known interval of difference between the categories. In this case, it might have been less confusing if the number of children in each family was recorded and a percentage of those families having one, two, three, and so forth, were reported. In the case of the gender of the children, the

genders of the children in all the families could have been recorded and the percents of male and female children representing all the families could have been reported.

2. In this example, the sports information director forgot that, in most cases, the GPA originates from ordinal data where typically an A = excellent, a B = good, a C = average, a D = below average, and F = failure. The assignment of numerical values only helps confuse the issue. The ordinal data was reported as if it were interval or ratio data with known uniform degrees of difference between the points or the scale.

CHAPTER

3

# Samples and Populations

*"Dewey Defeats Truman"*

***Chicago Daily Tribune*, 1948**
**(Front-page headline)**

Statistical methods are used to help us make decisions. These decisions are made by reasoning deductively or inductively. In either case inferences are made.

When deductive reasoning occurs, we reason from what we know about the general nature of all things to help us understand the nature of a specific thing. By definition, deduction requires that the general nature of all things be known before an inference to the specific case can be made. For most of us, the expenditure of time, energy, and money involved with the collection of data concerning the general nature of all like things can become prohibitive. In some situations, however, the needed data concerning the general nature of all things may be readily available. In these situations, deduction to the specific is highly economical and accurate.

When inductive inference is made, we learn what we can about a few specific cases and then make an inference to all such cases. By definition, induction requires that a relatively small group of cases be considered and used to make inferences about the nature of all such cases. Making inductive inferences often requires less exhaustive data collection efforts because only a relatively few specific cases are considered. Inductive inference, however, is always associated with some degree of error because we are required to make inferences from a *few* cases to the larger set of *all* such cases.

Making inferences to help make decisions is one important component of statistics; but, so too is description. Before inferences can be made with a reasonable degree of accuracy, we must describe the available information. Four key concepts are important in the understanding of descriptive and inferential statistical methods: the population, the parameter, the sample, and the statistic.

## THE POPULATION

A population represents the totality of things with which we are concerned. A definition of a **population,** as it is used by the statistician, is the set of all things (persons, objects, events) of a specific description. For example, we can arbitrarily define some populations as all heart attack victims; or all people who registered to vote and actually voted in the last election; or all second-year medical students with B.S. degrees in engineering; or all left-handed, female, philosophy majors receiving academic scholarships at four-year, accredited colleges and universities in the U.S. These defined populations share three common characteristics.

First, each has been arbitrarily defined. A population can be very large or very small in size, very broad or very narrow in scope, depending on the arbitrary nature of its definition. All the rocks on the beaches of the Atlantic Ocean would represent a very large population of rocks. All the rocks examined by students in Dr. Smith's geology seminar would represent a very small population of rocks by comparison.

Second, the definition of a population indicates a specificity of description. This means that the members of a population must have at least one characteristic in common.

In the first population defined above, the common characteristic was a history of heart attack. All heart attack victims would constitute a very large population. In this case, all people would be grouped as either heart attack victims or non-heart attack victims.

| | | |
|---|---|---|
| Common Characteristic | All heart attack victims | All those not having had a heart attack |

Furthermore, although the definition of a population requires at least *one* specific common characteristic, it is important to note that the number of common characteristics can be increased at will. In the second population defined above, we have added a second specific characteristic. The first common characteristic was having registered to vote. The second was actually having voted in the last election.

| | **All Potential Voters** | |
|---|---|---|
| 1st Common → Characteristic | All those registered to vote | All those not registered to vote |

| | **All Potential Voters Registered to Vote** | |
|---|---|---|
| 2nd Common → Characteristic | All those who voted | All those not voting |

As you have probably noticed, with the addition of each arbitrarily defined specific common characteristic, the number of potential members of the population is reduced. In the third population, the first common characteristic is enrollment in medical school, the second is second-year status, and the third is having earned a B.S. degree in engineering.

| | **All Students** | |
|---|---|---|
| 1st Common → Characteristic | Enrolled in medical school | Not enrolled in medical school |

| | **All Students Enrolled in Medical School** | |
|---|---|---|
| 2nd Common → Characteristic | Second-year students | Not second-year students |

| | **All Second-year Students Enrolled in Medical School** | |
|---|---|---|
| 3rd Common → Characteristic | Earned B.S. in engineering | Without B.S. in engineering |

Adding many specific common characteristics can severely limit the size of the population which you have defined. In the last population defined above, the actual number of left-handed, female philosophy majors receiving academic scholarships at four-year, accredited colleges and universities in the U.S. may be only a few. Note that, as the population defined becomes more limited, so too does the inference made to that population. For example, the results of an election that was determined by the votes recorded by all registered voters *who voted* should not be used to describe the potential results of an election based only on the party affiliations of those voters who registered. Those who registered and voted may represent a smaller group than simply the group of registered voters. Some registered voters may not vote and those who do may switch affiliation.

Third, the key word in describing a population is "all." Populations are either finite, if there is a known upper numerical limit of qualifying units, or infinite, if there is no known limit on the number of units that qualify.

In the case of Dr. Smith's geology seminar, the population of rocks examined during the seminar would probably be known or easily counted and would, therefore, constitute a finite population. *All* the rocks examined in Dr. Smith's geology seminar would make up this finite population. Furthermore, it would be possible to collect and study *all* of the rocks examined during Dr. Smith's seminar. After that, we could make accurate generalizations about the quantity and quality of the rocks Dr. Smith used during his seminar. In this case, a small, finite population was defined which made it possible to collect and analyze all the members of the population.

In the case of the infinite population of *all* the rocks on beaches of the Atlantic Ocean, there is no known upper limit of units that qualify as rocks. With infinite populations, such as this, we are often unable to collect and analyze all qualifying units of the defined population because of the constraints of time, money, access, and man power. When we are faced with the problem of trying to learn what we can about infinite populations—the nature of all things—we can select a sample from the population and base our inferences about the nature of the population on what we have learned about the sample.

## THE SAMPLE

When the data are collected from only a fraction of the total group, we are working with a **sample** of the population. If we were interested in studying all living heart attack victims to determine what role perceived stress might have in the determination of subsequent clinical episodes of heart attack, we could select 500, 250, 125, or even 5 subjects who qualified as heart attack victims to form our sample. If we were able to somehow involve 99% of all living heart attack victims in our study we would still only have a sample of all living heart attack victims. Not until we were able to involve 100% of all living heart attack victims in our study could we claim to have measured the role of perceived stress among the population of living heart attack victims. Anything less than 100% of the total population is a sample.

The results of a study with a sample consisting of 99% of the population would usually be more representative of the population than would the results of a study based on a sample involving only 10% of the population. Likewise, a study of a sample of 500 heart attack victims would often provide more dependable results than would a study that involved only 5 victims of heart attack. The rule to remember here is that *usually* the larger the sample, the better off we are when we attempt to generalize as to the nature of the entire population. This rule is not always true. As you might have guessed, how we select the sample is also very important. The results of a study of 500 living heart attack victims selected from the membership lists of a weight reduction program might provide information completely different from the results of a study of 50 living heart attack victims selected to represent both genders, several races, and a variety of socioeconomic conditions.

In most research situations we cannot hope to be able to collect all the members of a population, let alone measure them. Therefore, we select a sample from the population and measure the members of the sample. We then use what we have learned from the sample to make inferences about the whole population.

## THE PARAMETER

In those relatively rare situations in which we can actually collect information from each and every member of a population, the resulting measures are referred to as **parameters.** In the case of Dr. Smith's geology seminar, after obtaining the weights of each and every rock, we could then calculate the average weight of the rocks that make up the population of rocks discussed in the seminar. This average weight would represent a parameter of this specific population.

Again, we must note that parameters are very difficult to determine when the size of the population approaches infinity. It would be virtually impossible, for example, to calculate any parameter associated with the population of all rocks on the beaches of the Atlantic. One thing to remember is that if parameters can be determined, deduction becomes highly accurate.

## THE STATISTIC

A **statistic** results from the measurement of a sample. If, for example, we collected 10 rocks from a beach on the Atlantic Ocean and determined their average weight, this weight would constitute a statistic. Since we usually have much easier access to samples than to populations, statistics are much easier to calculate than are parameters.

As in the sample of 10 rocks collected from the Atlantic beach, we can use the statistic generated (average weight of the rocks) to make inferences concerning the average weight of all rocks on the beaches of the Atlantic. As you might guess, the inference made from our sample of 10 rocks concerning the average weight of all rocks on Atlantic beaches will probably not be completely accurate. This error of prediction may be reduced by collecting larger samples, and/or improving our sample selection methods. Yet, this is exactly the method of inferential statistics. That is, predicting unknown population parameters from known sample statistics—inductive reasoning.

## METHODS OF COLLECTING SAMPLES FROM POPULATIONS

The ultimate goal of sampling is to select a bias-free sample from a defined population. **Bias** is a lack of sample representativeness. That is, a biased sample does not fairly represent all members of the population and results in sample measures (statistics) that differ systematically from the population measures (parameters). Therefore, biased samples result in inaccurate inferences. Consider the following example of sample bias from the realm of political polling.

During the depression year 1936, a presidential preference poll was conducted by *The Literary Digest.* The candidates were Alf Landon, Republican, and Franklin D. Roosevelt, Democrat. A sample of over 10 million potential voters—derived from various telephone books and from a list of names provided by state motor vehicle registries—was selected from a list of prospective subscribers to the digest. Ballots were sent to members of the sample and fewer than 25% were returned (approximately 2.5 million). The results of the poll were calculated and Landon received 57% of the vote to a mere 40% for Roosevelt. The digest published the results: Landon in a landslide. Unfortunately for the digest and for Alf Landon, the results of the election that year found Roosevelt to be the landslide winner with over 60% of the votes. Following the election, *The Literary Digest* went out of business.

What happened? Clearly, Republicans were significantly overrepresented in the sample. In 1936, those voters who could afford to buy magazines and own telephones and automobiles (Republicans) were not representative of the voting majority. Also, as is the case with mail-in polls, the sample selected itself. Clearly, not everyone who was mailed a ballot returned it, and those who did were perhaps more motivated to make their choice known than was the population in general.

Bias, a systematic, directional sampling error, was responsible for the inaccurate inference and for the demise of *The Literary Digest.*

## Representative Samples

Samples that fairly represent all members of the population from which they are collected offer the best opportunity for accurate inferences. Various methods are employed for the selection of samples from populations. It is important to note that a poorly specified population promotes inaccurate sampling and the introduction of bias, for, without a clearly defined population, we are unable to effectively determine the membership of any sample taken from that population. The inclusion of members in the sample that are not part of the population and/or the exclusion of members from the sample that are included in the population will surely introduce bias. Once the population of interest has been clearly defined there are several sampling methods that may limit bias.

## Simple Random Samples

Perhaps the best method for limiting bias is the method of **random sampling.** This method meets many of the important assumptions underlying a variety of statistical procedures and it is one of the most important techniques used

**TABLE 3.1** Random Numbers

| | |
|---|---|
| 88608 | 10834 |
| 40122 | 07973 |
| 42374 | 45446 |
| 01272 | 95673 |
| 51274 | 05845 |
| 26299 | 19663 |

in statistics. It serves as the foundation for all sampling methods, no matter how sophisticated. Additionally, random sampling enables us to estimate sampling error.

A random sample constitutes a subset of a population selected in such a way that each population member has an equal chance of being selected. One of the best ways of selecting a random sample involves the use of a random number table. Random number tables are available in many statistical texts. These tables are simply lists of members generated to ensure that each number has equal representation. When you plan to use a random number table to help you select a random sample from a population, only three simple steps are needed: first, clearly define the population of interest; second, assign consecutive numbers to each member of the population; and third, use the table to select the random sample.

A column of numbers taken from a random number table has been presented in Table 3.1.

For illustration, assume we have a clearly defined population of blood counts for 95 subjects. The next step is to number each of the subjects from 01 to 95. Now imagine that we wanted to select a random sample of 10 of these 95 population members. If the random number table has two pages of numbers you could flip a coin to select one page or the other—heads for page one, or tails for page two. The selection of a place to start on one of the pages could be done by simply closing your eyes and touching a spot on the page with your pencil. You should record your starting point in case you want to verify your selection later. Imagine also that we have previously decided to proceed down the columns of numbers. You could equally as well proceed across from right to left, but do not change your direction during selection. Now, pretend your pencil stopped on the top line of the random number table (3.1). As we proceed down the table, reading the first two numbers in each row, we select the following 10 sample members:

| | |
|---|---|
| Case No. 88 | Case No. 26 |
| Case No. 40 | Case No. 10 |
| Case No. 42 | Case No. 07 |
| Case No. 01 | Case No. 45 |
| Case No. 51 | Case No. 95 |

Why did we select two-digit numbers? Because our sampling frame is identified by two-digit numbers, 01 to 95. What if we were to find a duplication? In application, the duplicate is simply ignored. What if we encounter number 96? Again, the sampling frame only defines the numbers 01 to 95. Disregard all numbers outside the defined sampling frame. Please try to keep in mind that the 10 numbers selected only identify which of the subjects, 01 to 95, have been randomly selected; the sample itself consists of the set of blood counts.

In some situations it may be impossible for you to number the members of an entire population. For example, a complete list of all persons in a city who have AIDS or who use tranquilizers may not be available. In these and other similar situations you may be required to utilize whatever lists are available to you and randomly select from those.

## Convenience Samples

**Convenience samples** are often called accidental samples. These samples are often self-selected and are convenient, yet haphazard, collections of observations that provide little valuable information concerning populations. Since these samples seldom represent any population, they often are difficult to analyze and interpret. Information gained from phone-in and write-in surveys, man-on-the-street polls, and surveys of college classes usually cannot be generalized beyond such groups without risking serious errors of inference. Television and magazine advertisements often report product information that has been generated from samples without known populations. For example, when you read or hear, ''88% of those interviewed picked Mousebites Cat Treats over other leading brands,'' keep your checkbook in your pocket until you know who was interviewed and what brands were compared.

## Systematic Random Samples

Systematic sampling, unlike convenience sampling, is an acceptable sampling technique. Like simple random samples, **systematic random samples** serve to fairly represent all members of the population from which they are selected. Inferences from systematic samples are considered to be more accurate than the results from simple random samples. The difficulty with systematic sampling is that there is no known statistical theory that can be used to indicate the degree of error associated with the inferences made. In practice, sampling error is estimated as if the samples were random; thus the resulting estimates of error are slightly more conservative.

When do we use systematic sampling? Systematic samples are frequently used, when a complete list of the cases constituting the population (a sampling frame) is available, to select a sample of a certain size or proportion from a population. For example, suppose we wished to select a 20% sample of the blood counts of the 95 subjects we discussed earlier. Twenty percent is equal to 20/100, or 1/5. We could start by selecting a number between 1 and

5 randomly, using a random number table. We would select our sample by picking that subject (1 to 5) and, systematically, every fifth subject thereafter from the numbered list of 95 subjects. If the randomly selected number between 1 and 5 was 2, we would then select subject number 2, subject number 7, subject number 12, subject number 17 . . . subject number 92.

Our complete set of systematically selected subjects according to case number would include:

| **Subject No.** | **Subject No.** | **Subject No.** | **Subject No.** |
|---|---|---|---|
| 2 | 27 | 52 | 77 |
| 7 | 32 | 57 | 82 |
| 12 | 37 | 62 | 87 |
| 17 | 42 | 67 | 92 |
| 22 | 47 | 72 | |

This 20% systematic sample results in 19 subjects being selected from the sampling frame of the population of blood counts of 95 subjects. Hence, 19/95 = 0.20 = 20%.

Be familiar with your population before you decide to systematically select samples. If the data are arranged in a cyclic fashion, systematic sampling can result in the selection of a nonrepresentative sample. Consider the following:

If laboratory rats were housed by gender in individual, alternate, numbered cages, a cyclic situation with regard to gender would result:

| | | | |
|---|---|---|---|
| Cage 1 | Cage 2 | Cage 3 | Cage 4 |
| Rat 1 | Rat 2 | Rat 3 | Rat 4 |
| **Female** | **Male** | **Female** | **Male** |

If we then systematically sampled fractions of 1/2, 1/4, 1/6, 1/8, etc., we could produce a nonrepresentative sample having either no male or no female rats, thus introducing gender bias.

With this one exception, systematic sampling, like simple random sampling, can produce samples that are not biased and are representative (within a given margin of sampling error) to the population from which they were selected.

## Stratified Random Samples

When we know something about the important characteristics of a defined population we may wish to select a sample that represents these characteristics according to their proportion in the population. To accomplish this, we must first learn how each characteristic is represented proportionately in the population (strata) and then sample randomly and proportionately from each strata (subgroup) of the population.

We will again consider the population of 95 blood counts for this illustration. Suppose we learn that 38 of these subjects are females and 57 are males and that avoiding gender bias in our sample selection is important. We can then calculate that our population includes 40% female blood counts and 60% male blood counts (38 females/95 subjects = 0.40 = 40% females and 100% − 40% = 60% males). Next we must ensure that our sample includes 40% randomly selected females and 60% randomly selected males. There are many ways to accomplish this gender breakdown in the sample. One method is to divide the population into two groups, males and females, and then randomly select two females for every three males using a random number table until the sample is complete (40 females:60 males = two females:three males). This would result in a selected sample representing the proportional gender breakdown (2:3) of the population.

**Stratified random sampling** is useful when bias may creep into the results of a study because of proportional differences in the characteristics of the population. Population characteristics such as age, gender, race, religious preference, socioeconomic status, and voting affiliation are often stratified in order to ensure proportional sample representation.

### Cluster Samples

The use of **cluster sampling** is of value when simple random sampling techniques become impractical or uneconomical due to the size and nature of the population defined. Suppose we are interested in the opinion of a population of persons in a certain large city concerning neighborhood crime. Instead of attempting to use simple random sampling techniques we could randomly select groups or clusters of people, as defined by neighborhoods, or plots of two-block-by-two-block areas within all the residential areas of the city.

For example, the residential areas of the city could be divided into plots of similar population density or area, which could then be numbered. Next, a sample of several plots could be randomly selected. Within each randomly selected plot, each resident could be interviewed, or a secondary, simple random, systematic or stratified sample selection could be made.

## SAMPLING ERROR

**Sampling error** can be defined as the difference between our sample and the population when we select a random sample to represent a defined population. Sampling error is always expected, since it is unlikely that the sample will exactly represent the population of interest. Whenever we sample, we must remember that no matter how thoughtful we have been in our selection process, the sample results will not be completely accurate.

With this in mind, we will use a hypothetical situation to illustrate the idea of sampling error. Imagine that we are able to obtain a total serum cholesterol value for each adult American. Since we have cholesterol values for all American adults, we can claim to have the population of total serum cholesterol values for adult Americans. From these, we compute the average total serum cholesterol value for the population and determine that the population average for all adult Americans is 220 units of serum cholesterol. Assuming there are no measurement errors (in blood analysis, equipment calibration, recording, etc.), the true average total serum cholesterol level for the population is 220 units.

Next, suppose we randomly select a sample of 1,000 adults from the population, measure their total serum cholesterol levels, and find the sample's average to be 230 units. Suppose we then randomly selected another sample, and another, and reported sample averages of 215 units and 225 units, respectively. Although our sample averages are not exactly equal to the population average, they are relatively close. In this example, the variation of the sample averages from the population average (disregarding measurement errors) is due to sampling error. Moreover, these sample deviations from the true (hypothetical) population average are normally expected. It is important to note that sampling error is not a mistake, such as making an arithmetic error; it is *an expected amount of deviation* due to chance alone. Additionally, with the selection of repeated samples of the same size, we would expect approximately the same degree of sample deviation positively (above the population average) as negatively (below the population average)!

What do you think the result would be if we selected 25 samples of 1,000 adult Americans and for each calculated a sample average, and then calculated a grand average using the 25 sample averages? The result would be that the grand average of the 25 sample means would be very close to the population average.

## SUMMARY

In 1948, the Republicans chose Thomas E. Dewey and the Democrats nominated Harry S. Truman as candidates for president of the United States. In September, six weeks before the election, the American Institute of Public Opinion, headed by George Gallup, went to press and predicted a win for Dewey. The morning after election day, a Chicago newspaper, *The Daily Tribune,* printed a headline proclaiming Dewey's victory just before the final vote count was complete. The next day the votes were counted. Truman won.

After some study, two mistakes seemed to be at the source of Gallup's faulty inference. First, he polled too soon. Many people who were polled had not made up their minds and others would change their minds before the election, which was six weeks away. Second, most of the people polled who claimed to be undecided probably voted for Truman. When undecided responses were made, Gallup placed them in a pile and divided them equally between the candidates.

Clearly, the collective mind of the sample George Gallup polled six weeks before the election was not representative of the collective mind of the population who voted on election day. Perhaps, if the election had been held six weeks earlier, Dewey might have won. As was obvious after the vote count, the sample taken six weeks earlier did not represent the voting population.

## FOR REVIEW

1. The population is the set of all things having at least one characteristic in common.
2. A parameter is any numerical value describing a population.
3. A sample is a subset of a population.
4. A statistic is any numerical value describing a sample.
5. Bias occurs whenever a sample systematically misrepresents a population.
6. A simple random sample results when a subset of a population is selected in such a way to ensure that each member of the defined population has an equal chance for selection into the subset.
7. Convenience samples are collections of observations that seldom represent a clearly defined population.
8. Systematic samples are samples that are selected by repeating a predetermined sequence of selection, usually from a randomly selected starting point. A defined population could be sampled by selecting every fifth case, for example. Systematic samples are useful when the sampling frame is known and noncyclic in nature.
9. Stratified samples are samples that duplicate the known proportions of cases within certain strata (subgroups) of the defined population.
10. Cluster samples are samples that consist of selected groups (plots) of cases within the defined population. They are often used to sample very large or infinite populations.
11. Sampling error is the amount of deviation resulting from chance differences between parameters of the defined population and statistics of a random sample of that population.
12. Measurement error results from mistakes made during the collection, analysis, interpretation, and presentation of data.

**Chapter Knowledge and Comprehension Application**

Which of the following are parameters and which are statistics? Why?

1. The average blood pressures for the 25 members of the first-year medical school class at the local medical center.
2. The average weight of 50 members of a particular corporate weight-loss program.
3. The voting preferences of 100,000 Americans collected by random telephone interviews.
4. The average resting heart rate of the members of the 1992 U.S. Olympic wrestling team.

**Answers**

1. Parameter: The average blood pressures were determined by measurement of *all* 25 members of the defined population.
2. Statistic: The average weight was determined by measurement of a subset of 50 of the members of a population. The total number of members is not reported.

3. Statistic: The preferences only represent 100,000 Americans. It is not clear what the population might be (all Americans with phones who were willing to respond?).
4. Parameter: The average resting heart rate represents the population of *all* members of the defined population.

Draw a random sample of 4 students from the following set of 12 (See Table 3.1; start with number 88608.):

| | | | | | |
|---|---|---|---|---|---|
| Joe | Bob | Al | Phil | Bill | Paul |
| Mary | Ann | Sue | Jill | Barb | Alice |

**Answers**

*Step 1* Define the population to be sampled: a set of 12 student names.

*Step 2* Number each member:

| | | | |
|---|---|---|---|
| 01 Joe | 04 Ann | 07 Phil | 10 Barb |
| 02 Mary | 05 Al | 08 Jill | 11 Paul |
| 03 Bob | 06 Sue | 09 Bill | 12 Alice |

*Step 3* Use the random number table. Select:

01 Joe
10 Barb
07 Phil
05 Al

CHAPTER

4

# Searching for Answers: Experimentation and Observation

*"We don't know a millionth of one percent about anything."*

THOMAS ALVA EDISON

As we search the literature for answers to questions, we encounter data that have originated from previously conducted studies and other sources. When we read and interpret this information, or conduct studies of our own, it is helpful to understand how traditional sources of information have been categorized and to understand the strengths and weaknesses inherent in each category.

In this chapter we will discuss three traditional sources of data: those categorized as opinions, experiments, and observational studies. The discussion that follows will help you become a more critical reviewer of the literature and will help you design and conduct better studies of your own.

## OPINIONS

An **opinion** constitutes a point of view expressed by a particular individual or organization. Opinions may or may not be based upon a large body of experience and/or knowledge.

We can observe opinions, at one level, by reading the captions of cartoons in the newspaper; and at another level, by reading exhaustive literature reviews in highly acclaimed professional research journals. The opinion, whether in the form of cartoon captions, man-on-the-street interviews, editorial comments, or lengthy reviews of literature, can provide the reader with another, often provocative, point of view. In some cases, opinions such as editorial comments and literature reviews not only serve you by providing interesting interpretations of data, but can provide you with additional bibliographic information as well.

Opinions in the form of literature reviews may or may not include the most recent, the most germane, or the most popular information available. The authors of opinions often *select* certain experiments and observational studies for review in order to support their own particular points of view.

For example, if we were interested in reviewing the data available concerning abortion, we might begin by reading literature reviews of the subject. Imagine that we selected two reviews to read. The first review selected was written by the editor of a pro-life journal, and the second review was written by the editor of a pro-choice journal. As you might guess, the two editors probably would *select* different studies for review in order to support the positions taken by their respective journals. If the two editors had selected a few common studies for review, the interpretation of these studies would probably be dramatically different. What we have learned by reading *both* reviews is something about *both* of the opinions held about abortion. It would be a mistake to limit our exposure to only one journal's opinion if our purpose was to learn about both points of view.

Opinions can provide sources for continued study and can offer new or different interpretations of data for our consideration. Yet an opinion, no matter how qualified, no matter how provocative, no matter how popular, is still, after all, just another point of view.

## EXPERIMENTAL STUDIES

Another traditional source of information is the experimental study. In contrast to opinions, experimental studies are designed to control the influence of factors that reduce objectivity. In fact, *experimenter control* is its most distinguishable characteristic. In **experimental studies,** the researcher controls the study's membership, the factors to which the study's membership is exposed (treatments), and the conditions under which the study is conducted. The ways in which the study's membership is assigned to different treatments, together with any relationships there may be among the various treatments, constitutes the experimental study's *design*. We will briefly describe four of the more common experimental designs.

The simplest model of an experimental study is the **controlled clinical trial.** The controlled clinical trial design is used to determine the effect of a particular treatment method. Controlled clinical trials have traditionally used people as experimental subjects.

In the basic form of this design, subjects are *randomly assigned* to one of two comparable groups. For example, if the treatment of interest involves the use of a new drug in arthritis therapy, the two groups would include the group of subjects which receives the new drug therapy (**treatment group**) and the group of subjects which receives the **placebo** or the standard therapy (**control group**). It is very important to make reasonably sure that the two groups are comparable before we assess the effect of the new drug therapy. If, for example, the two groups were not represented equally in terms of severity of the disease or type of previous therapy, the effect of the new drug might be confused with differences that existed between the treatment and control groups *prior* to the treatment (selection bias). To ensure that the two groups are reasonably comparable, the groups are randomly selected from the same population of subjects and randomly assigned to a group. This *random group assignment* affords each subject an equal chance of being a member of either the treatment group or the control group. In the controlled clinical trial, the investigator controls the selection of subjects and the assignment of subjects to treatment groups.

Not only must the investigator guard against **selection bias,** but he or she must also exert controls to avoid investigator bias and subject bias. **Investigator bias** may cloud the results of the study if the investigator has knowledge of the treatment and the control group memberships. This knowledge may subconsciously influence the investigator's interpretation of the results or measurement of responses to the treatment. **Subject bias** may occur when the subjects themselves know whether they are receiving the treatment or the placebo (an inert or innocuous substance used instead of the treatment by the control group). This knowledge alone may be blamed for differences between the groups after treatment. The best way to avoid investigator and subject bias is to conduct the controlled clinical trial in a **double-blind** fashion. That is, the study is conducted in such a way that neither the investigator nor the subjects know to which group the subjects have been assigned. This requires that the members of the treatment and control groups be assigned and coded by a third party who only reports group membership *after* the data have been collected. To help avoid subject bias, it is important for the placebo to be administered to the control group members in the same fashion as the drug or treatment is administered to the treatment group members.

Before we go on to briefly review other experimental study designs it may be helpful to discuss some terms that are often used to describe these designs.

Treatments are frequently called **independent variables.** Independent variables are manipulated by the researcher while observing their effects on the response or **dependent variables.** In our example of the controlled clinical trial, the independent variable was the new drug therapy. The dependent variable or response variable could have been some indicator of joint inflammation caused by arthritis. If the new drug was an effective treatment, the members of the treatment group would probably have responded with less joint inflammation than did the members of the control group who received the placebo or standard therapy. If the new drug could be tolerated without undue risk or increased side effects, it may be considered more effective than the placebo or the standard therapy.

When the researcher is interested in simultaneously determining the effectiveness of multiple treatments, often the **completely randomized design** is used. In this design each case (or subject) has the same chance of being assigned to any treatment being considered. Therefore, if four treatments were being investigated, each case would have a one-in-four (1/4) chance of being assigned to any one of the four treatment groups. In the example of the controlled clinical trial, subjects were randomly assigned to either the control group or the new drug therapy (treatment) group. In that case, a particular subject's chances of being selected as a control or treatment group member was one in two or 1/2.

When the researcher is interested in using a nominal or ordinal variable to block or divide the study into two or more sets before randomly assigning control and treatment group membership, the **randomized blocks design** may be used. If, for example, we learn that the etiology of arthritis may be different for women than for men, we may decide to block our cases according to gender—males or females. Once we have separated our cases (subjects) by gender into two blocks we can randomly assign control and treatment group membership for the males and for the females, respectively. If one gender responds to the treatment differently, we can essentially block out the gender effect. If the differences in drug response are similar among the males and among the females, we could then obtain an overall comparison of the drug's effect that is not obscured by gender.

Another design, the **split-plot design,** involves assigning cases (''whole plots'') randomly to treatment groups and then assigning subsets (''split plots'') to a second kind of treatment group. This design originated from agricultural experiments in which the cases were plots in a field. To illustrate this design, imagine that our arthritis subjects (''whole plots'') are randomly assigned to two drug therapy treatment groups. Imagine also that both of these drugs, although used to reduce arthritis inflammation, also causes a side effect: deafness. We could then randomly assign each subject's ears (the ''split plots'') to one of two procedures to control or reduce the degree of deafness.

There are many types of experimental designs, each demonstrating appropriate controls and randomization. The reasons for such a variety of experimental study designs are to avoid bias and ensure experimental objectivity while conserving the researcher's time, money, and energy.

Experimental methods can, in many situations, help the researcher answer questions. However, in some situations the experimental study is not practical. For example, imagine that we are interested in determining the effect that smoking three packs of cigarettes a day during pregnancy has on the health and well-being of a mother and her newborn at delivery. We could design a controlled clinical trial in which a population of healthy women, upon first detection of pregnancy, were randomly assigned to either a control group or a smoking group. The women randomly assigned to receive the smoking treatment would be asked to begin smoking three packs of cigarettes a day and to continue smoking three packs each day for the duration of the pregnancy. As you might guess, we would find that few, if any, of the expectant mothers would cooperate with such a foolish request. As in this example and others like it, the experimental study is simply not a practical or ethical method of investigation.

In situations such as this we often must rely on epidemiological methods of investigation. **Epidemiological studies** are used to determine how and why diseases or conditions that contribute to disease are distributed in populations. The most common approach used to determine both the prevalence of a disease condition and the characteristics of the members of the population with the disease involves methods of observation.

## OBSERVATIONAL STUDIES

**Observational studies** are those in which the researcher observes the members of a particular sample of a population. Two general approaches are used; each provide different epidemiological sampling methods for the determination of disease cause. These two observational approaches are commonly known as prospective (cohort) and retrospective (case-control) studies.

**Prospective** or **cohort studies** are observational studies in which a sample of healthy subjects are identified and observed into the future (prospectively). It is important to note the sampling method. In prospective studies, samples are designed and selected to ensure that the suspected cause of the disease condition (e.g., smoking behavior) is present in one subset of the sample and absent in the other subset. The two subsets of the sample are called cohorts. Cohorts share common characteristics (e.g., age, gender, race, etc.) except for the suspected cause of the disease condition. The reason the prospective study's sampling design separates the sample into cohort groups according to the *presence or absence* of the *suspected cause* of the disease is simply because the sample consists of healthy subjects who have not yet developed the disease or condition of interest. After prospective observation, the cohort subsets of the sample are compared in terms of the proportion of each subset who develop the disease or condition (e.g., maternal and newborn disease states).

Prospective observational studies have several *disadvantages*. First, they are time-consuming. Often, members of cohort groups must be followed for months or years before the disease state occurs. Second, they are expensive. Since prospective studies follow the cohort groups indefinitely, the financial burden associated with man-hours spent in data collection, testing, and record keeping is often prohibitive. Third, prospective studies have not proved to be successful in studying diseases or conditions with low population incidences.

Prospective observational studies, however, have several important *advantages*. Most important, the prospective study permits accurate, relatively unbiased estimates of disease incidence and causes in a population. This is primarily because data is collected for specific reasons, under uniform conditions; and because other relevant variables such as age, gender, and occupation may be included.

Unfortunately, prospective observational studies are not able to *prove* a causal relationship between the suspected cause of a disease or condition and the occurrence of the disease or condition. The reason for this drawback is that the effects of possibly confounding variables are not accounted for by randomly assigning cohort groups to the treatment (e.g., smoking behavior) or to the control (e.g., nonsmoking behavior). For example, those who smoke have been shown to drink more alcohol and coffee when compared with nonsmokers. Therefore, the disease may have been induced not by smoking, but by alcohol or coffee consumption. To claim that smoking behavior *causes* unhealthy moms and newborns at delivery may not, therefore, be completely true. The occurrence of the disease or condition could be *confounded* by other variables associated with the suspected cause.

**Retrospective** or **case-control studies** collect past information from cases (individuals with a specific disease or condition) and controls (those without the disease or condition) to identify differences in exposure to the suspected cause of the disease or condition. For example, when comparing smoking habits during pregnancy and the health of mothers and newborns at delivery, the researcher could identify women who had delivered *unhealthy* newborns (cases) and women who had delivered *healthy* newborns (controls). Once these cases and controls had been identified, the researcher would seek to determine if the cases differed from the controls in terms of the proportion of each group who had demonstrated smoking behavior during the pregnancy.

In retrospective observational studies the researcher begins by identifying cases and controls related to the disease or condition and then looks back in time to identify the suspected cause.

Retrospective studies have a major weakness. Often because of incomplete records or the necessity of relying on the memories of the subjects, it is difficult to select appropriate controls. Therefore, the two groups (cases and controls) may not be comparable at all. This absence of demonstrated comparability between cases and controls may result in bias and confounding.

The advantages of the retrospective study are that they are economical and are particularly applicable to infrequent conditions or rare diseases.

## SUMMARY

Opinions, experiments, and observational studies are standard techniques used for providing information to help us answer questions. The questions that we frame determine the design we choose as well as the methods of description and/or analysis of the data provided.

The data provided by opinions, experiments, and observations may be used descriptively. That is, each may provide data that can be used to estimate characteristics of a population. These estimates (based on the sample used) are only as accurate as the methods used for sampling (e.g., the population of interest). Therefore, it is critically important to clearly understand not only the descriptive estimates generated by opinions, experiments, and observations, but also the ways in which these estimates were gathered. We must continually be on guard against failures in the methods used to determine if the estimates generated have been biased or somehow confounded by the effects of uncontrolled variables.

The data provided by experimental studies and prospective (cohort) observational studies can be used descriptively as well as analytically. These analytical efforts seek to determine the degree of association between a variable of interest and a factor in the population. For example, the estimates of systolic and diastolic blood pressures could serve to *describe* the blood pressure status of a certain population of subjects; while the response of the subjects (in terms of changes in blood pressure status) when treated with a certain drug could be used to analyze the degree of association between the drug and the estimated changes in the population's blood pressure status. Again, it is important to remember that the reported analytical information is subject to the design weaknesses of the study. Again, we must ask ourselves if the samples used were comparable, if they fairly represented the population of interest, and if the study's design protected against bias.

To best answer questions, researchers use statistical methods and attempt to design the most efficient and unbiased investigations possible. The data provided are then described, analyzed, interpreted, and presented to others so that decisions can be made. Students not only need a practical critical review of the scientific literature in their field, but also must select the study design that best fits their needs when conducting research.

## FOR REVIEW

1. An opinion is a point of view expressed by an individual or an organization.
2. An experimental study involves a plan or design in which the experimenter imposes control over the selection of cases and the methods, treatments, or conditions to which the cases are exposed.
3. An observational study involves a plan or design in which observations are made of events or conditions over which few, if any, controls are imposed.
4. The controlled clinical trial is a carefully designed experimental study often conducted using a sample of human subjects to determine the effectiveness of a particular treatment.
5. The treatment (experimental) group comprises the group of cases randomly assigned to receive exposure to a particular condition or treatment. The treatment variable is the independent variable and the variable over which the experimenter can impose control.

6. The control group consists of the group of cases randomly assigned to receive the placebo or alternate therapy.
7. The independent variable is the treatment controlled by the researcher.
8. The dependent variable is the response variable over which the researcher has no control. Dependent variables comprise the subjective responses to the independent or treatment variable.
9. The placebo is the inert or innocuous substance or condition to which the randomly selected members of the control group are exposed in place of the treatment.
10. Selection bias results when samples are selected from different populations to constitute the membership of the control and treatment groups. After treatment, differences in the responses of the groups are confused by preexisting differences in the respective samples. Selection bias causes doubt that the treatment was totally responsible for the subsequent differences between the control and treatment groups.
11. Subject bias results from the participant's knowledge of their membership in the control or the treatment group, and his or her perception of the research objectives.
12. Investigator bias occurs, potentially, when the investigator is aware of the membership of each case in either the control or treatment group. The investigator may allow this knowledge to subconsciously influence his or her evaluation or interpretation of the study's results.
13. The double-blind experiment is one in which neither the subject nor the investigator knows to which group (control or treatment) the subject has been randomly assigned. Double-blind experiments avoid subject and investigator bias.
14. A completely randomized experimental study is one in which each case has an equal chance of being assigned to the control group or to any treatment group.
15. A randomized blocks design is an experimental study design in which cases are first divided into mutually exclusive blocks and then randomly assigned to the control and treatment groups separately within each block.
16. The split-plot design is an experimental design in which cases ("whole plots") are randomly assigned to treatment groups and then subunits of each case ("split plots") are randomly assigned to a second treatment group.
17. Prospective (cohort) observational studies are those in which comparable groups of healthy subjects are identified and observed for the onset of a response (disease or condition) into the future.
18. Retrospective (case-control) observational studies are those in which cases (those expressing a characteristic of interest, e.g., a disease) and controls (a comparable group that does not express the characteristic of interest) are identified and past information is collected concerning exposures to the suspected cause of the characteristic.

**TABLE 4.1** Frequencies of Middle-Aged, White, Professional Males Grouped by Disease (with and without CAD) and Level of Dietary Saturated Fat

| | Dietary Saturated Fat Levels High | Low |
|---|---|---|
| Group 1 (with CAD) | 23 | 7 |
| Group 2 (without CAD) | 14 | 16 |

**TABLE 4.2** Frequencies of Middle-Aged, White, Professional Males Grouped by Blood Pressure Status (Group 1 = Hypertensives; Group 2 = Normals) and Drug Treatment Group

| | **Blood Pressure Groups** | | | |
|---|---|---|---|---|
| | *Group 1 High (n= 30)* | | *Group 2 Normal (n= 30)* | |
| *Post-Treatment Blood Pressures* | placebo n = 15 | new drug n = 15 | placebo n = 15 | new drug n = 15 |
| *High* | 13 | 7 | 2 | 0 |
| *Normal* | 2 | 8 | 13 | 15 |

**Chapter Knowledge and Comprehension Application**

Which of the following are or have resulted from opinions, experimental studies, or observational studies? Explain the reasons for your decisions. Interpret the results of each study.

1. Consider the following study design (Table 4.1) and the resultant data: A study was conducted to determine the effect of high and low levels of dietary saturated fats on cardiovascular disease among middle-aged, white, professional males. Group 1 consisted of 30 subjects demonstrating clinical evidence of coronary artery disease (CAD) and Group 2 consisted of 30 subjects demonstrating no clinical evidence of coronary artery disease.
2. Consider the following study design (Table 4.2) and the resultant data: A study was conducted to determine the effect of a new drug on the blood pressures of a group of 60 middle-aged, white, professional males. Group 1 consisted of those males with clinical evidence of high blood pressure and Group 2 consisted of those males with clinical evidence of normal blood pressure. Placebo and new drug therapies were randomly assigned.

**TABLE 4.3** The Reported Five-year Survival Percentages of Coronary Heart Disease Patients Grouped by Therapy

| Group | Five-Year Survival Rate |
|---|---|
| Group 1<br>Surgical Intervention<br>(5 studies reported) | 60% |
| Group 2<br>Drug and/or Life-style Alteration Therapy<br>(5 studies reported) | 80% |

3. Consider the following study design (Table 4.3) and the resultant data: Research was conducted by the *Big Canyon Wellness Journal* to determine the effect of surgical intervention as opposed to drug and/or life-style alteration therapy on the survival of victims of coronary heart disease. Ten studies were reported—five involving surgery and five involving drug and/or life-style alteration. The results were presented as the percentage of each group who survived five years after treatment.

**Answers**

1. This is a retrospective observational study because the cases (those clinically defined as coronary heart disease victims) and the controls (those without clinical evidence of disease) were identified and their dietary habits with regard to saturated fat consumption were compared. We must assume that the dietary habits that existed *before* they were clinically identified were compared. The results seem to indicate that diets described as high in saturated fat contributed to the disease. However, the results from Group 2 seem to suggest, at least, that dietary saturated fat may not be the only reason for the disease. The two groups seem to be reasonably comparable—middle-aged, white, professional males. Yet, perhaps the study was confounded by the presence or absence of other factors such as dietary cholesterol or levels of exercise.
2. This is an experimental study using a randomized blocks design. Subjects were randomly assigned to the placebo and new drug treatment groups after they were blocked according to blood pressure status (high or normal). The placebo effect was similar in both blocks, yet the treatment (the new drug) reduced the blood pressures for over half of the subjects who received the drug and had high pressures. Even though the new drug did not bring about an increase in pressure among those with normal pressures who received the drug, it might be of interest to know if pressures fell in this group.
3. This is an opinion. There is no evidence presented to indicate that any of the subjects in any of the studies were comparable. It would be good to read each of the studies cited and form your own opinion.

CHAPTER

**5**

# Describing Data

*". . . Lake Wobegon where all the women are strong, all the men are good-looking, and all the children are above average."*

**Courtesy of GARRISON KEILLOR.**

We are all experts at describing data. We have been describing data informally all of our lives. For example, we are frequently asked to describe a movie, a restaurant, or the weekend's activities. When asked to describe a movie, we are not expected and do not attempt to give intricate, scene-by-scene details. Instead we usually try to summarize the movie in terms of its action, romance, comedy, or sound track. Perhaps we compare aspects of the movie to other movies we have seen.

Whenever we attempt to describe a set of available information, whether scenes from a movie or data from a study, we must choose descriptors that meet our purposes. With movie description we may need only indicate our opinion of the movie within some frame of reference. We may need only categorize the movie as worth the price of admission or not. Or, we may decide to rank the movie from 1 to 10, where 1 represents the rank of the worst movie of all time and 10 represents the all-time best.

The description of study data has a single most important purpose: the promotion of a better understanding of *the data at hand.* Inferential statements are not included in the description of data; instead the emphasis is on the organization, summarization, and abbreviated symbolic description of the available sets of data.

Imagine that we were interested in describing a single variable collected from a large sample of cases. For example, say we were interested in describing the age at which a sample of 1,000 women received their hysterectomies. In this example, the age of each woman at the time she underwent a hysterectomy is the variable of interest. The sample consists of 1,000 independent cases (women).

One method to describe the ages of the sample at the time they received hysterectomies would be to group the women by age category (younger, middle, and older) and then report the *ratio* of the number of cases in each age group relative to the total group (n = 1,000). A second method of description would be to *graphically display* the number of women (out of 1,000) at each age who received a hysterectomy. A third method could include reporting both the age representing the middle of the distribution of 1,000 ages when the ages were arranged from youngest to oldest—the *median* age—and the difference between the youngest and oldest age, or the *range* of the ages. Each of these three examples of data description could serve to enhance our understanding of the age at which these 1,000 cases underwent hysterectomies. Moreover, each of the previous examples represents one of three general classifications of data description:

1. Ratios, proportions, and rates
2. Graphs and tables
3. Central tendency and dispersion

Before we begin our discussion of these classifications it is important to briefly mention that data descriptions are also influenced by whether the data are discrete or continuous in nature.

## DISCRETE AND CONTINUOUS DATA

Data may be expressed as discrete or continuous units. If the data occur as integers (e.g., whole numbers: 1, 2, 3, . . . ) the data are **discrete.** The number of visits to the dentist's office, the number of patients in a hospital unit, the number of family members, and the number of reportable accidents are examples of discrete data. Counts of nominally and ordinally scaled data are also considered discrete.

If data can reasonably take on fractional values, they are considered **continuous.** For example, measures of a subject's weight could result in any value within a range of relevant values. We could measure a subject's weight as 150 pounds, 150.3 pounds, or 150.33 pounds. For convenience, continuous data is often rounded to the nearest relevant value. Interval scaling and ratio scaling result in continuous data.

When describing data it is important to consider the nature of the data. While it is reasonable to report the number of children in each family, a problem is created when we report that, on average, there are 2.3 children per family.

## RATIOS, PROPORTIONS, AND RATES

Ratios, proportions, and rates are used by allied-health researchers to compare and contrast vital demographic and morbidity data. Population demographic data—including indicators of age, gender, race, and occupation—are most commonly provided by a census of the population. Data pertaining to vital events (such as births, deaths, marriages, and divorces) are required by law to be registered. Disease prevalence data (morbidity data) are reported by morbidity surveys.

Ratios, proportions, and rates use counts of categorized (age, gender, race) cases, which are discrete units, for relative comparison.

### Ratios

Ratios are perhaps the least complicated form of relative measurement. A definition follows:

$$\textbf{Ratio} = \frac{\text{the number of cases experiencing a particular event during a particular time period}}{\text{the number of cases experiencing a different event during the same time period}}$$

When using ratios for relative comparison two requirements must be met. Both the numerator and the denominator *must* use the *same units of measurement,* and the data must be collected and compared over the *same period of time*.

For example, suppose we were interested in the gender ratio of live births in a particular county during the month of August last year. Suppose we found that, of 132 live births, 72 were females and 60 were males. The ratio of female live births to male live births in the county during August of last year was $\frac{\text{72 females}}{\text{60 males}}$ or 1.2:1.

In summary, the population of live births (the count of live births) for the county during August of last year was divided into two parts (by gender). A live birth was the same unit of measurement in both the numerator and the denominator, and the data were collected over the same time period (August, last year).

## Proportions

Proportions and percentages are each created by dividing a fraction of a population by the total population. A definition for the **proportion** of cases experiencing a particular event follows:

$$\textbf{Proportion} = \frac{\text{the number of cases experiencing a particular event during a particular time period}}{\text{the total number of cases at risk of experiencing the particular event during the same time period}}$$

Using the previous example, the proportion of male live births in the county during August of last year would be:

$$\frac{60}{60 + 72} = \frac{60}{132} = .045$$

When a proportion is multiplied by 100 we have a percentage. The percentage of male live births in the county during August of last year would be:

$$\frac{60}{60 + 72} = \frac{60}{132} \times 100 = 45\%$$

It is important to remember that, like ratios, proportions and percentages require that the numerator and the denominator consist of the same units collected over the same time interval. Also, just as proportions *may* be multiplied by 100 to form percentages, ratios and proportions *may* be multiplied by other constants such as 1,000, 10,000, or 100,000.

## Rates

**Rates,** unlike ratios and proportions, have numerators that are accumulated over a period of time (commonly a year); their denominators are fixed points, estimating the population (total number of cases) at risk at one point in time. A definition for the rate of those cases experiencing a particular event follows:

$$\textbf{Rate} = \frac{\text{the number of cases experiencing a particular event during a particular time period}}{\text{the total number of cases at risk of experiencing the particular event, at a particular time, during that time period}}$$

Rates, like ratios and proportions, are also multiplied by constants such as 100, 1,000, 10,000, or 100,000 for convenience.

**TABLE 5.1** Presence and Absence of Disease: Two Groups of Cases

| Groups | Present | Absent | Total |
|---|---|---|---|
| 1 | 10 | 10 | 20 |
| 2 | 10 | 90 | 100 |

The types of rates commonly used are crude, specific, and standardized.

**Crude rates** are those computed for the entire population, disregarding specific categories such as gender and age. For example, the annual crude death rate is the number of deaths in a calendar year divided by the population on the first of July of that year, multiplied by 1,000.

**Specific rates** account for existing differences within such categories as age and gender. For example, the age-specific death rate is the number of deaths in a specific age group in a certain calendar year, divided by the population of the same age group on the first day of July of that year, multiplied by 1,000.

**Standardized rates** are used to make fair comparisons between two or more groups possessing different distributions (for example, age and gender).

When comparing counts of cases experiencing a particular event in two distinct groups, the actual counts are difficult to interpret unless each group contains the same number of cases. Often a relative comparison is made. That is, the actual counts are converted into a ratio, a proportion, a percentage, or a rate, and then these relative measures are compared. Consider the information in Table 5.1. Here we are interested in comparing the counts of cases experiencing a particular disease in two distinct groups. As indicated by the counts of cases representing the two groups in Table 5.1, each group reported 10 cases having the disease. Is this a good comparison of the groups?

The answer is no. A much better comparison is to report the number of cases with the disease in each group, relative to the size of the group, and then compare these relative values. To demonstrate this, we first determine the relative values for each group by computing the percentage having the disease. For example, Group 1 had 20 members, 10 of which had the disease: $10/20 = .5 \times 100 = 50\%$. Fifty percent of the members of Group 1 had the disease. Group 2 had 100 members, 10 of which had the disease: $10/100 = .1 \times 100 = 10\%$. Ten percent of the members of Group 2 had the disease. These relative values (50% with the disease in Group 1, and 10% with the disease in Group 2), when compared, help us better understand the relationship between the groups in terms of the presence of the disease. These relative values are less likely to misrepresent the data than would the simple counts alone: 10 cases in Group 1 and 10 cases in Group 2. Ratios, proportions, percentages, and/or rates help to represent each group relative to its size when comparing groups of unequal size.

## TABLES AND GRAPHS

Tables and graphs are popularly used to organize and summarize data and descriptions of data. Virtually any set of numerical data can be presented in tabular or graphic fashion. When used properly, tables and graphs present the summarized data in a way that is easily understood. When used improperly, tables and graphs present the summarized data in a way that is confusing or even misleading. There seems to be no single set of instructions to follow when creating a table or a graph for each and every possible set of data. When constructing tables and graphs, remember to clearly label them; in this way the table or graph can be understood without additional supportive explanations. A properly constructed and labeled table or graph explains itself.

Even though thoughtfully constructed and clearly labeled tables and graphs are less likely to distort the data, distortions may still occur. When large sets of data are summarized and presented in tabular or graphic format there is ample opportunity for mistakes and the subsequent misrepresentation of the data.

When biased or otherwise inaccurate data are summarized in a well-designed and correctly labeled table or graph, it is nearly impossible to identify errors unless the source of the data summarized is cited. If the source of the data is not cited clearly in the table or graph, beware! If the source of the data is cited, this is still no guarantee that the information presented in the table or graph is without error or bias. However, in this case you have the option of critically reviewing the source of the data presented.

Sometimes accurate data are summarized in misleading ways. For example, the slope of a graph may be adjusted by manipulating the scales on the graph's vertical and/or horizontal axes. In Figure 5.1, you can see how the graph on the left can be changed to the graph on the right by simply expanding the scale on the vertical axis (percentage) and shortening the scale on the horizontal axis (years).

It is important to note that both graphs present the data accurately; however, the impression given by the graph on the left is dramatically different than the impression given by the graph on the right. Since there is no ''correct'' scale for any line graph, we must be careful to recognize the effect of a particular scale used on the overall impression we receive.

Sometimes only selected portions of the data are summarized in a table or graph. This selective description of the data can result in the misinterpretation of the data set. Look at the graphs presented in Figure 5.2. The graph on the left summarizes the percentage of employees participating in a health-promotion program for each month during 1994. The graph on the right summarizes the percentage of cases participating in the program in April, May, June, July, and August of 1994. The graph on the right presents only a selected portion of the data for the year. The impression given by the graph on the right has also been influenced by the manipulation of the scales on the vertical and horizontal axes. These graphs present dramatically different impressions

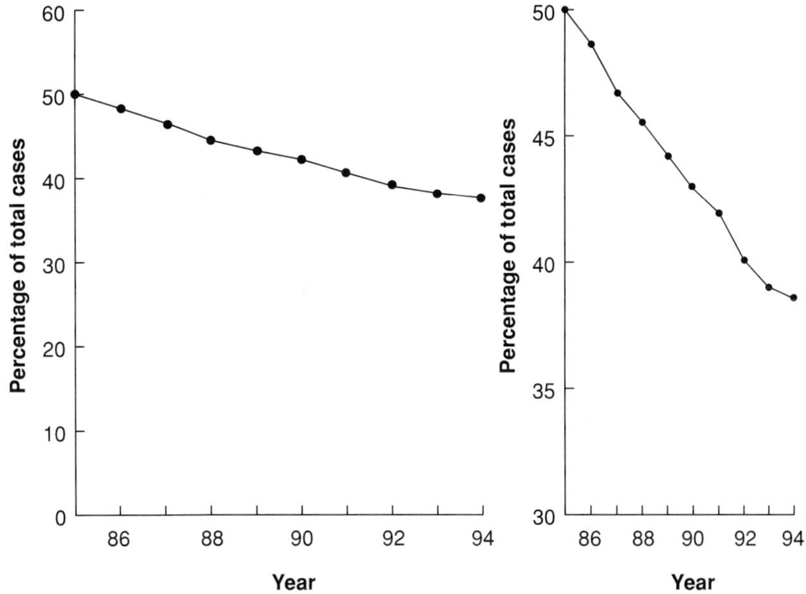


**FIGURE 5.1**
Two line graphs depicting the same data: percentage of total cases for each year, 1985–1994.

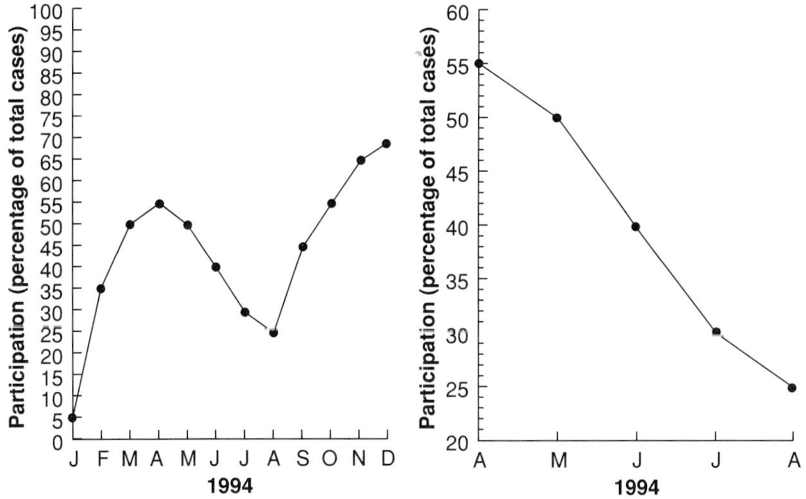


**FIGURE 5.2**
Two line graphs depicting the percentage of total employees participating in the company's health-promotion program for 1994.

**TABLE 5.2** Format of a Typical Table

| **Row Headings, Subheadings** | **Column Headings, Subheadings** |
|---|---|
| | contents of table |

**TABLE 5.3** Blood Alcohol Concentration Percentage (BAC% for Each of 30 Cases)

| | | | | |
|---|---|---|---|---|
| 0.00 | 0.00 | 0.00 | 0.01 | 0.01 |
| 0.01 | 0.02 | 0.02 | 0.03 | 0.03 |
| 0.03 | 0.04 | 0.05 | 0.05 | 0.06 |
| 0.07 | 0.08 | 0.08 | 0.09 | 0.09 |
| 0.10 | 0.10 | 0.11 | 0.13 | 0.13 |
| 0.14 | 0.15 | 0.16 | 0.17 | 0.18 |

of the data. When you encounter tables or graphs that present only selected portions of the data, you should be cautious in your interpretation. Find out what was omitted and why.

Since tables and graphs offer such authoritative first impressions, we must guard against being fooled, and critically evaluate the methods used to collect and present the data.

## Tables

Data and descriptive summaries of data are presented in **tables.** Tables should be concise. An example of the format generally followed in constructing tables appears in Table 5.2. Each table is numbered to differentiate it from other tables; the number is followed by the title of the table. Row and column headings and subheadings are used to clearly label the contents of the table.

An uncomplicated table, one that simply lists data, is illustrated in Table 5.3. In this table there is no need for row or column headings since the 30 blood alcohol concentration (BAC percentage) values were organized for convenient presentation in five columns, each having six BAC% values. The rows and columns have no special significance other than listing the values in order, from lowest to highest.

In Table 5.4 the BAC% values presented in Table 5.3 have been grouped into one of four intervals. The frequency, relative frequency, and relative frequency percentage for the BAC% values have been calculated for each interval. This type of table is very successful in summarizing large data sets. Table 5.4 provides us with more information than a simple list of ordered values. For example, from Table 5.4 we can see that 12 of the 30 cases had

**TABLE 5.4** Frequency, Relative Frequency, and Relative Frequency Percentages of Blood Alcohol Concentration (BAC% presented in Table 5.3)

| Interval (BAC%) | Frequency (Counts) | Relative Frequency (Count/Total) | Relative Frequency % (Rel. freq. × 100) |
|---|---|---|---|
| 0.00–0.04 | 12 | 12/30 = .40 | .40 × 100 = 40% |
| 0.05–0.09 | 8 | 8/30 = .27 | .27 × 100 = 27% |
| 0.10–0.14 | 6 | 6/30 = .20 | .20 × 100 = 20% |
| 0.15–0.19 | 4 | 4/30 = .13 | .13 × 100 = 13% |
| Total | 30 | 30/30 = 1.00 | 1.00 × 100 = 100% |

BAC% values between 0.00 and 0.04. These 12 cases together represent four-tenths, or 40%, of the total group.

As we have demonstrated, tables can successfully contribute to an improved understanding of the data. Such tables have several common characteristics:

1. The sources of the data used are clearly cited. Only original data are used without citation.
2. The title of the table is concise and is preceded by a reference number. It clearly indicates what data are tabulated and how these data are measured. When appropriate, the number of cases represented by the data is reported.
3. The table does not contribute to its own misinterpretation. All symbols and abbreviations are explained. The table's rows and columns are clearly labeled. Row, column, and grand totals are reported when appropriate.

The final form of a table ultimately depends upon the type of data summarized and the goals of the table's author. Remember, a table should be able to stand alone. That is, it should be easily understood without the need for additional explanation from the supporting text.

## Graphs

**Graphs** are also used to summarize data. Like tables, graphs should be clear, concise, and appropriately labeled. Unlike tables, graphs present powerful visual generalizations of the data at a glance. Tables usually require a brief period of study before we are able to ascertain the meaning of their contents. With graphs we are immediately influenced; since they provide such persuasive impressions of the data, it is of utmost importance to avoid misinterpretation. Graphs that effectively promote our understanding of the data have several common characteristics:

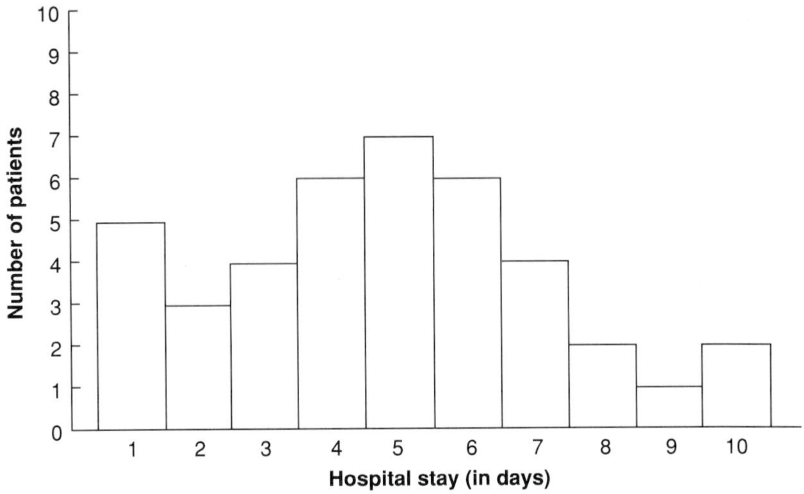


**FIGURE 5.3**
Histogram showing duration of hospital stay for 40 randomly selected patients.

1. The sources of the data used in the graph are clearly cited. Only original data are used without citation.
2. The title of the graph is preceded by a figure number for reference. The graph's figure number and title usually appear below the graph. The title of the graph is clear, concise, and descriptive.
3. Traditionally, graphs are read from the left to the right and from the bottom to the top. All legends, scales, scale labels, and keys are therefore placed accordingly. The variable of interest is traditionally scaled along the horizontal axis (also called the **X axis** or **abscissa**). Time (minutes, days, months, or years) usually appears on the horizontal axis as well. The vertical axis (also called the **Y axis** or **ordinate**) usually displays counts, for example, the number of cases or the percentage of the total cases. The scale used on the horizontal axis has equal intervals of measurement. The same is true for the scale used on the vertical axis. A scale is selected for the vertical axis so that the zero point is apparent.

Numerous methods for graphing data have been developed. Some of the more popular methods have been named and are discussed below.

## *Histograms*

**Histograms** are an easy way to graphically depict counts of data measured on an interval scale. In Figure 5.3 a histogram is used to graph the length of hospital stay, in days, for 40 randomly selected patients. The vertical axis is scaled as the number of patients (counts, frequency) and the horizontal axis is scaled for length of stay with one day as the interval. The selection of a

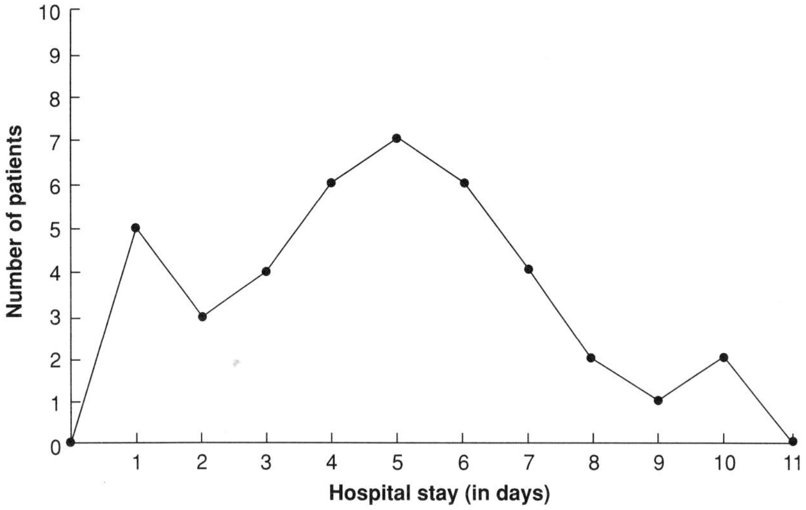


**FIGURE 5.4**
Frequency polygon showing duration of hospital stay for 40 randomly selected patients.

particular interval width is, to some extent, arbitrary. In this case the practical interval width of one day is used. Widths of 1, 2, 3, 5, 10, and 20 units are commonly used. Note in this example that the scales on each axis are uniform and the vertical scale begins at zero. The histogram allows us to see that 5 of the 40 patients stayed one day, 3 of the 40 patients stayed 9 days or more, and 19 of the 40 patients stayed 4 to 7 days. The histogram is one of the most commonly encountered graphs. The histogram can also be used to compare two sets of data.

When comparing two sets of data, both sets can be graphed on the same histogram. If the two sets of data have *equal* numbers of cases, then frequencies, relative frequencies, or percentages can be used to construct the vertical axis. Care should be taken to ensure that both sets of data have been graphed using the same horizontal axis intervals. When the two sets of data have *unequal* numbers of cases they may still appear together in the same histogram if they are compared using their relative frequencies or relative frequency percentages on the vertical axis. Histograms usually present no more than two sets of data for one figure. If more than two sets of data are to be presented at once for the same figure, a frequency polygon is typically used.

## *Frequency Polygons*

**Frequency polygons** are simply line graphs used with ordinal, interval, or ratio data to display the frequency of cases (vertical axis) for some occurrence (horizontal axis). Frequency polygons, like frequency histograms, are not used with nominal (categorical) data. An example of a frequency polygon using the data presented as a histogram in Figure 5.3 appears in Figure 5.4. Here we

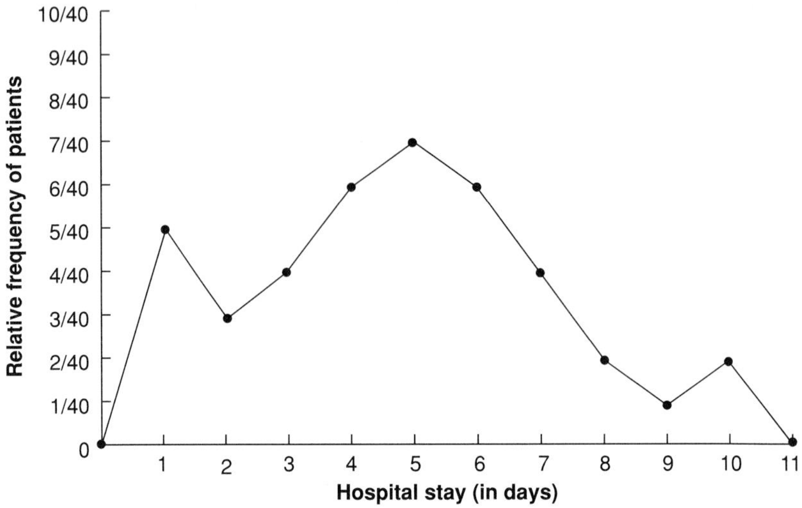


**FIGURE 5.5**
Relative frequency polygon showing duration of hospital stay for 40 randomly selected patients.

have indicated five patients for day 1, three patients for day 2, four patients for day 3, using dots. We then connected the dots to form a line.

**Relative frequency** and **relative frequency percentage polygons** are often developed to compare data sets of unequal sizes. Examples of these using the same data expressed in Figures 5.3 and 5.4 appear in Figures 5.5 and 5.6. As you can see, the shapes of the polygons are the same whether relative frequencies or relative frequency percentages are used.

Frequency polygons can serve to present a comparison of several sets of data. Suppose we were given access to data pertaining to the length of stay for three distinct groups of patients. These three groups represent 40 pediatric, 40 young adult, and 40 geriatric patients. In this case, we could compare the length of stay for all three groups simultaneously on the same figure by creating a polygon for each group. Again, if the groups were of *unequal* sizes, relative frequencies or relative frequency percentages could be used to account for the group size difference.

Sometimes we are interested in using simple line graphs to display **cumulative percentages.** These cumulative percentage curves are called **ogives.** If we plot the cumulative percent on the vertical axis and the length of stay on the horizontal axis for the data in Figure 5.6, we create the cumulative percentage curve in Figure 5.7. From this curve, for example, we ascertain that 95% of the cases stayed fewer than 10 days in the hospital. Also, 80% of the cases stayed longer than two days.

Frequency polygons and histograms are very popular methods of data description. In most situations, either serves well to display data. Remember, to avoid clutter and subsequent confusion, histograms are usually limited to

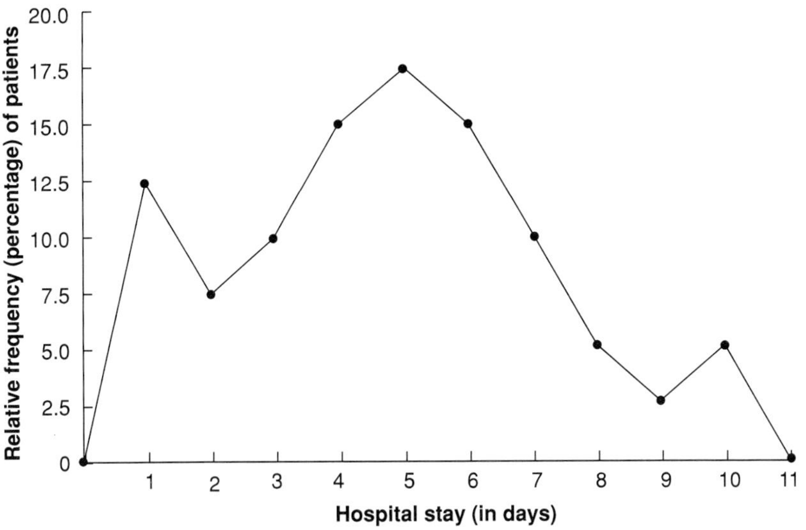


**FIGURE 5.6**
Relative frequency percentage polygon showing duration of hospital stay for 40 randomly selected patients.

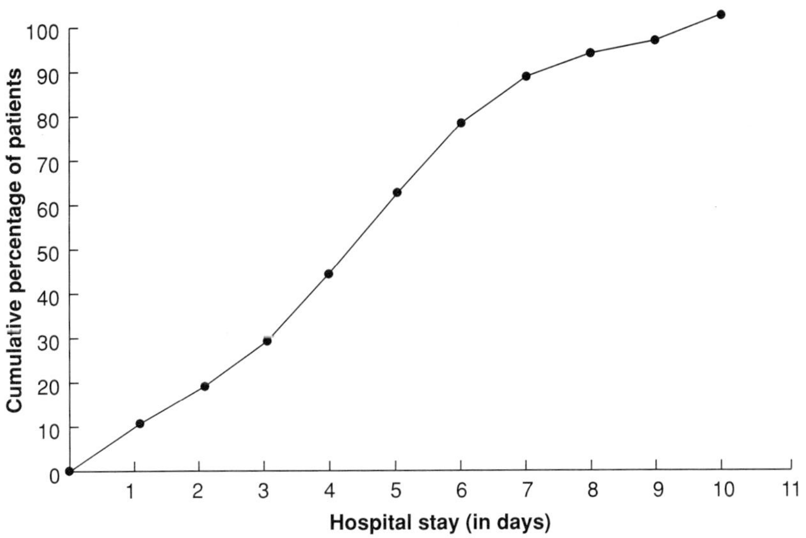


**FIGURE 5.7**
Ogive showing duration of hospital stay among 40 randomly selected patients.

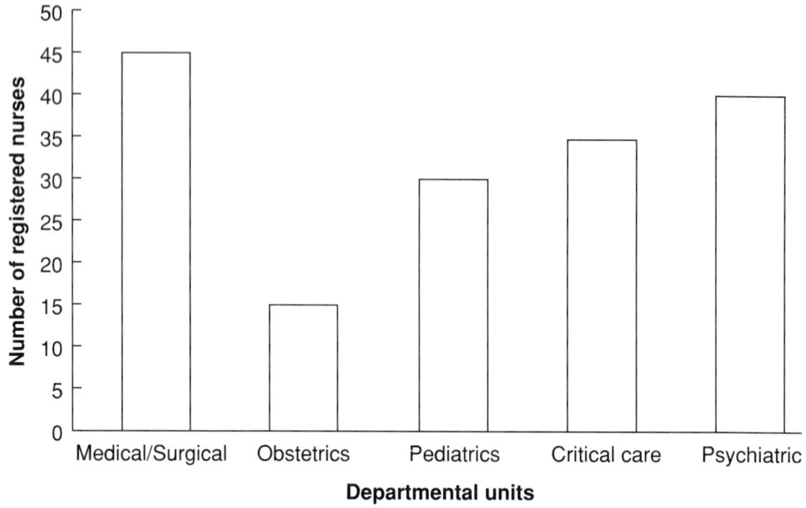


**FIGURE 5.8**
Bar graph of registered nurses by hospital departmental unit.

one or two sets of data on one figure. Polygons may be used more successfully to display two or more data sets at once. Only ordinal, interval, or ratio data are used with frequency polygons and histograms.

### *Bar Graphs*

The data presented on the horizontal axis of a histogram are assumed to be at least ordinal in nature. With bar graphs, this assumption does not hold. The data presented on a **bar graph** are nominal categories and do not have a known underlying order or interval.

In Figure 5.8 the horizontal axis depicts several hospital departmental units (categories) while the number of registered nurses (frequency) is plotted on the vertical axis. Notice that the bars indicating the number of registered nurses in each of the departmental categories do not touch. This indicates that the data presented on the horizontal axis represent distinct categories and do not suggest some underlying order or interval.

## Describing Shapes of Distributions

When the shapes of distributions are described, certain terms are commonly used to indicate how these distributions differ from "normal." In Chapter 7 we will continue our definition of the "normal distribution," but for now, let us define it as a bell-shaped, symmetrical distribution such as the one in Figure 5.9.

### *Skewness*

Distributions are described as being skewed when a greater number of cases are located toward one end of the distribution than the other, resulting in a loss of symmetry.

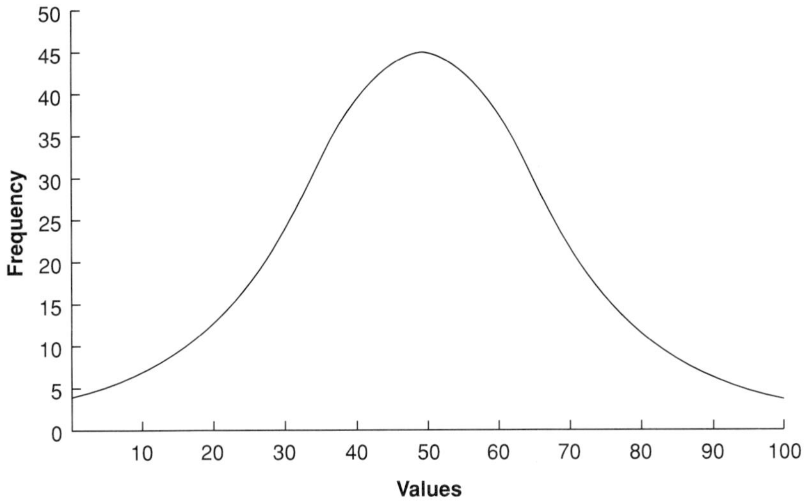


**FIGURE 5.9**
The normal distribution.

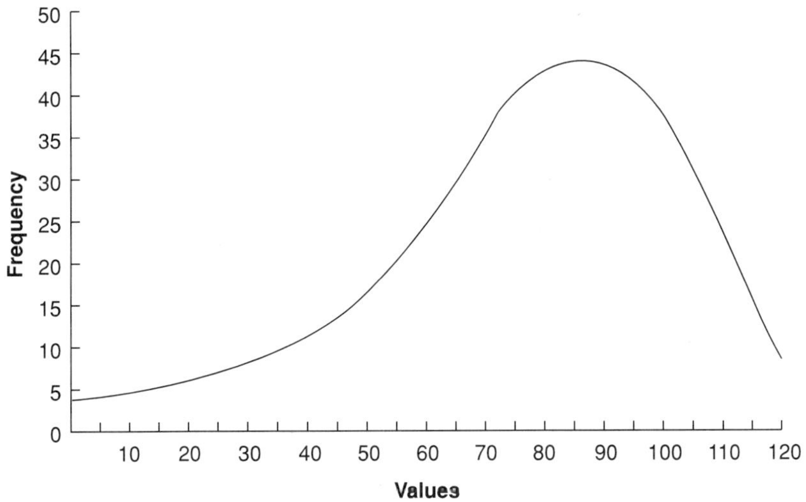


**FIGURE 5.10**
A negatively skewed distribution.

When most of the cases occur toward the positive or right end of the distribution with the tail tapering off toward the negative or left end (as in Figure 5.10), we have a **negatively skewed distribution.**

When most of the cases are clustered toward the negative or left end of the distribution with the tail tapering off toward the positive or right end (as in Figure 5.11), we have a **positively skewed distribution.** It is helpful to remember that the tail of the curve points in the direction of the skew.

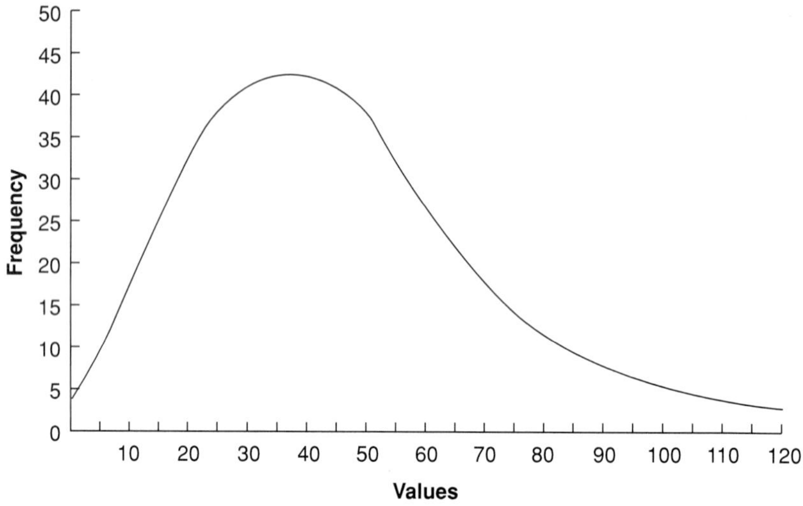


**FIGURE 5.11**
A positively skewed distribution.

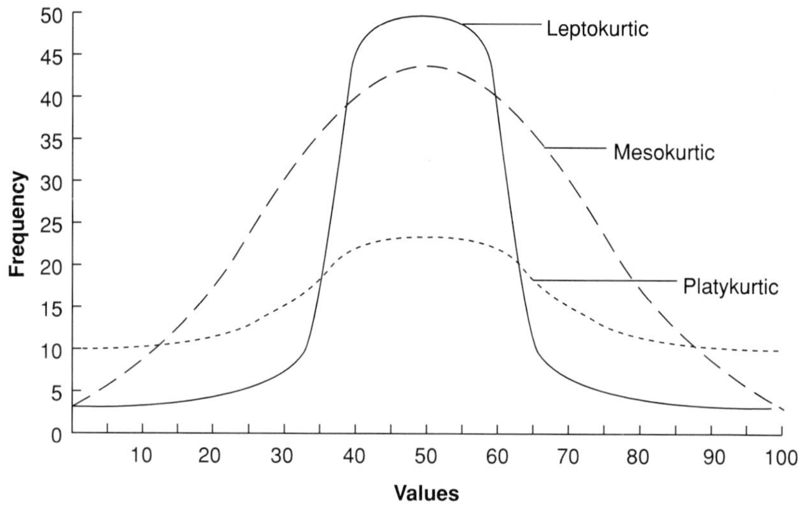


**FIGURE 5.12**
A comparison of three distributions in terms of their kurtosis.

## *Kurtosis*

When distributions are described as skewed, we know they are asymmetrical and not normally distributed. When distributions are symmetrical but still not normal, they are either too peaked or too flat. The term **kurtosis** refers to the peakedness or flatness of a distribution. A bell-shaped, symmetrical distribution (a normal distribution) is referred to as a **mesokurtic** distribution (see Figure 5.12). If the distribution is more peaked than the normal distribution it is called **leptokurtic,** while a flat distribution is **platykurtic.**

As we have seen, tables and graphs can be used to present information in a variety of ways. When information is presented in this form, always take time to question the construction of the table or graph. Ask yourself if the impression given by the table or graph is a function of the manner in which the data is presented rather than the actual differences in the data itself. That is, if the data were presented in a different manner, would they imply a different conclusion?

What we have learned with ratios, proportions, and rates is also important when evaluating tables and graphs. When comparing two distinct groups, make sure the groups are compared relative to their sizes. Also, if two tables or two graphs are compared, make sure they are constructed in the same way. With tables, make sure titles and column headings indicate a comparison of apples with apples. With graphs, make sure they have been constructed using the same units on the respective horizontal and vertical axes.

## CENTRAL TENDENCY AND DISPERSION

So far, we have discussed two general methods of data description. The first of these general methods includes the description of data using ratios, proportions, and rates; the second method includes the use of tables and graphs to visually describe the data at hand. Now we will investigate how statements of central tendency and statements of dispersion can be used together to help us describe the distribution of a set of data. Let us begin by discussing statements of central tendency and statements of dispersion separately; then we will discuss how they are used together to describe data.

### Statements of Central Tendency

**Statements of central tendency** are nothing more than attempts to describe the whole distribution of the data set by reporting one *most typical* value. These most typical values serve to represent the point or points about which most of the values in the distribution are centered. Three statements of central tendency are commonly used to describe data from health-related studies. These are the mean, the median, and the mode.

#### *The Mean*

The **mean** is the *average* value. The mean can only be used to describe interval and/or ratio data. Imagine that we have recorded the ages of 11 children with chicken pox. Suppose the ages were 2, 4, 5, 5, 6, 6, 6, 7, 7, 8, and 10 years. One way to represent these 11 ages with one most typical value is to calculate the mean age for the group of 11 children with chicken pox. To find the average age of this group we must first sum the ages ($2 + 4 + 5 + 5 + 6 + 6 + 6 + 7 + 7 + 8 + 10 = 66$), and then divide the total by the number of cases ($66 \div 11 = 6$ years). The mean age for this group is 6 years.

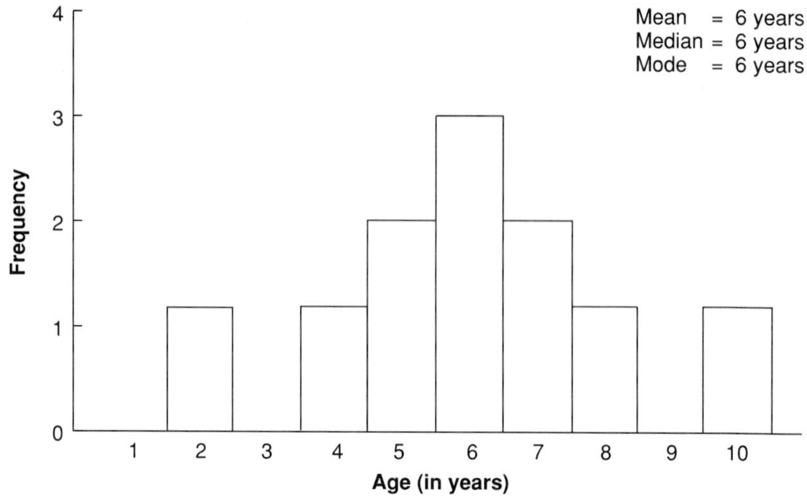


**FIGURE 5.13**
Frequency histogram of the ages of 11 children with chicken pox in Group 1.

### *The Median*

The **median** is the middle value in a group of ordered values. The median can be used to describe data measured on the ordinal, interval, or ratio scales. Look again at the ages of the 11 children with chicken pox: 2, 4, 5, 5, 6, 6, 6, 7, 7, 8, and 10. Notice that these ages have been ordered from the youngest to oldest, and the middle value is the sixth value from either end of the ordered distribution. The median age for this group is 6 years.

If we have an *even* number of values the median is found by averaging the *two middle values*. For example, imagine the ordered ages of a group of six children are 1, 2, 3, 4, 5, and 6 years. The two middle values are 3 and 4 years. The average of these is $(3 + 4 \div 2)$ 3.5. Therefore, 3.5 years represents the median value for this group of six ages.

### *The Mode*

The **mode** is the most frequently occurring value in a group of values. The mode can describe nominal, ordinal, interval, or ratio data. The most frequently occurring age in this group of 11 children with chicken pox is 6 years. Three of the children were 6 years old, and none of the other ages were represented more than twice. Therefore, 6 was the most frequently occurring age.

It is important to remember that there may not be a single, most frequently occurring value in the distribution, and if there is, it may not be unique. A particularly flat (platykurtic) distribution would probably not have a mode, and distributions with two or three modes are called bimodal or trimodal distributions, respectively.

As illustrated in Figure 5.13, the ages of the 11 children in Group 1 were distributed symmetrically around a mean of 6 years, a median of 6 years, and

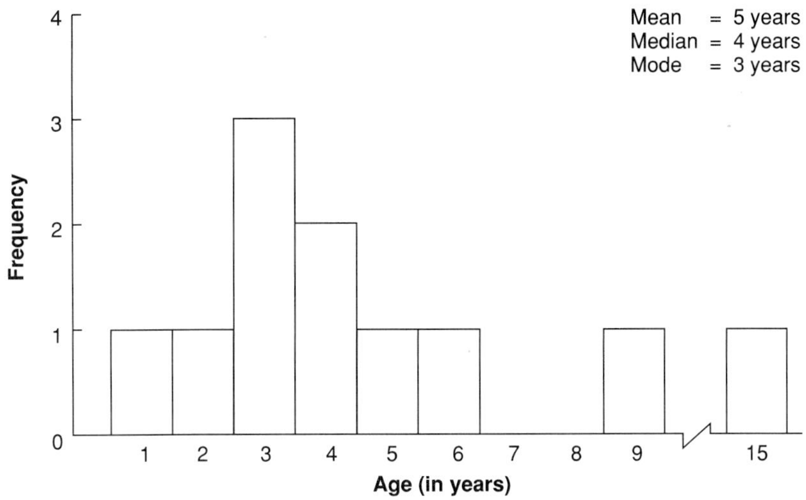


**FIGURE 5.14**
Frequency histogram of the ages of 11 children with chicken pox in Group 2.

a mode of 6 years. When the mean, the median, and the mode of a distribution are equal, the distribution is symmetrical.

In Figure 5.14 we have an asymmetrical distribution. Specifically, we have a positively skewed distribution of the ages of 11 children in Group 2. The mean of this distribution is 5 years. The median is 4 years and the mode is 3 years. With skewed distributions, the mean will be pulled toward the tail of the distribution, as in this case. The mode will always occur at the highest frequency point on the distribution, and the median will fall between the mode and the mean.

When we attempt to choose the most appropriate statement of central tendency to use when describing a set of data, two factors must be considered. First, as we have seen, the *shape of the distribution* influences our choice. If the distribution is **symmetrical,** the mean, median, and mode will be equal and each may be used as a most typical representative value. Leptokurtic distributions are often best described using the mode. When the distribution is **skewed** in either a positive or negative direction, the median is the best choice as a statement of central tendency.

The second factor to consider before choosing a statement of central tendency is the *scale of measurement* of the data being described. If the data are *nominally* scaled (categories) then our only choice is the mode. That is, we can indicate the most frequently occurring category with the mode.

When we have *ordinally* scaled data we may report the median or the mode. Only in the case of a very peaked distribution should the mode be selected. Otherwise, with ordinal data the median is reported.

With *interval* or *ratio* data the mean is reported if the data distribution is symmetrical and not exceptionally peaked. If the data distribution is skewed, the median is reported instead of the mean. Report only the mode if the distribution is exceptionally peaked.

## Statements of Dispersion

We have discussed three methods for describing the central, most typical value of a distribution of values. Yet, as we can see in Figure 5.12, several different distributions could all be described by the *same* central value. Each of the three completely *different* distributions in Figure 5.12 has the mean, median, and mode (a value of 50). To only report any one of these most typical values does nothing to help us understand the *spread* of the distribution of values. Therefore, to adequately describe any distribution we must not only report the appropriate statement of central tendency but also indicate how the values in the distribution are dispersed about that statement of central tendency. We will discuss five statements of dispersion: the range, the interquartile range, the average deviation, the variance, and the standard deviation.

### *The Range*

The least complicated indicator of dispersion is the **range.** The range is the difference between the highest and lowest values in a distribution of values. In Figure 5.13 we can see the highest of the values (ages) was 10 years and the lowest was two years. The range (10 − 2) or eight years, merely describes the span of the distribution of values. Knowledge of the range however, does not provide us with information concerning the nature of the values within the limits of the range. Another problem with the range is that it is influenced by extreme values in the distribution. Another statement of dispersion that is not affected by extreme values in a distribution is the interquartile range.

### *The Interquartile Range*

The **interquartile range** is the range of the values that constitutes the middle 50% of the values in a distribution. In Figure 5.15 we have two frequency distributions of values, each with an interquartile range of 6 to 9. Therefore, we know in each case the middle 50% of the distribution of values lies between 5 and 10. We also know that 25% of the values in the distribution of values lie above 9 and 25% lie below 6.

Also, in Figure 5.15 the upper distribution is symmetrical and has a range equal to 13, while the lower distribution is positively skewed and has a range equal to 20. This clearly indicates how extreme values influence the ranges of distributions. Note, however, that the interquartile range is the same for both distributions.

### *The Average Deviation*

Before we discuss the average deviation, we will explain what is meant by a **deviation.** Previously, when discussing the mean, we used the ages of a group of 11 children with chicken pox to calculate the group's mean. The ages of these children are 2, 4, 5, 5, 6, 6, 6, 7, 7, 8, and 10 years. The sum of the ages

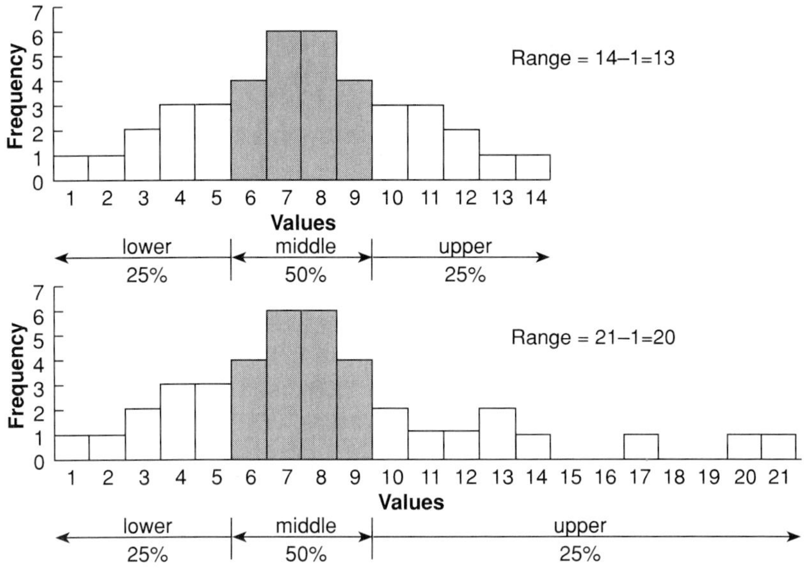


**FIGURE 5.15**
Frequency distributions of values with interquartile range of 5 to 9.

is 66, and the average age is 6 years (66 ÷ 11 = 6). We can now calculate a deviation from the average score (6 years) for each of the 11 children with chicken pox. A deviation is represented by the positive or negative difference between the mean of the group and each child's age. For example, in Table 5.5 we calculate the deviation between each child's age and the average age of the group.

Notice that the sum of the deviations from the mean is equal to zero. One of the properties of the mean or average is that it is the point in a distribution of measurements, such as these ages, about which the summed deviations are equal to zero. The symbol for summation is Σ (pronounced "sigma"). We could then write that the Σ of the children's ages is equal to 66 years; and the Σ of the deviations from the mean, $\Sigma(X - \overline{X})$, is equal to zero.

We know that a deviation represents the difference between a particular value and the mean of the group. A deviation can be found by subtracting the group's average value from any value in the group. Therefore, when we talk about deviations we imply deviations from the mean.

What then is the average or mean deviation? The **average deviation** is simply the average amount of absolute (positive and negative) deviation from the mean. The deviations of each of the values from the mean value are summed as if they were all positive, and this sum is divided by the number of values in the distribution.

The average deviation, although rarely used, represents a more refined measure of dispersion than does either the range or the interquartile range.

**TABLE 5.5** Deviations From the Mean Age For a Group of 11 Chicken Pox Sufferers

| Child's Age, X | Child's Age Minus the Mean ($\overline{X}$) | Deviations From the Mean, ($X - \overline{X}$) |
|---|---|---|
| 2 | 2–6 | −4 |
| 4 | 4–6 | −2 |
| 5 | 5–6 | −1 |
| 5 | 5–6 | −1 |
| 6 | 6–6 | 0 |
| 6 | 6–6 | 0 |
| 6 | 6–6 | 0 |
| 7 | 7–6 | 1 |
| 7 | 7–6 | 1 |
| 8 | 8–6 | 2 |
| 10 | 10–6 | 4 |
| Sum of the ages ($\Sigma X$) = 66 years | | Sum of the deviations ($\Sigma X - \overline{X}$) = 0 |

### *The Variance and the Standard Deviation*

The variance and the standard deviation are two commonly used statements of dispersion. The **variance** of a sample may be defined as the sum of the squared deviations from the mean value divided by the number of values minus one. The variance of a sample may be calculated by following these steps:

1. Find the deviation of each value in the set from the mean value.
2. Square each of these deviations.
3. Sum all of these squared deviations.
4. Divide the sum of the squared deviations from the mean by the number of values minus one ($n - 1$).

The **sample standard deviation** is the measure of dispersion or variability that is simply the square root of the variance.

We will use the data in Table 5.5 to demonstrate the calculation of the variance and the standard deviation. These data appear with some additions in Table 5.6. Notice that each child's age is represented by one of the 11 X values, $X_1 - X_{11}$. The sum of these X values is equal to a total of 66 years ($\Sigma X = 66$ years). The mean age of the sample is symbolized as $\overline{X}$ and is found by summing the X values and dividing by the number of X values.

Therefore:

$$\overline{X} = \frac{\Sigma X \text{ (sum of the X values)}}{n \text{ (the number of X values)}},$$

$$= \frac{66 \text{ years}}{11}$$

$$\overline{X} = 6 \text{ years}.$$

**TABLE 5.6** Squared Deviations From the Mean Age for a Sample of 11 Chicken Pox Sufferers

| **Child's Age (X) in Years** | **Child's Age (X) Minus The Mean Age ($\bar{X}$) in Years** | **$(X - \bar{X})^2$** |
|---|---|---|
| $X_1 = 2$ | $2 - 6 = -4$ | $(-4)^2 = 16$ |
| $X_2 = 4$ | $4 - 6 = -2$ | $(-2)^2 = 4$ |
| $X_3 = 5$ | $5 - 6 = -1$ | $(-1)^2 = 1$ |
| $X_4 = 5$ | $5 - 6 = -1$ | $(-1)^2 = 1$ |
| $X_5 = 6$ | $6 - 6 = 0$ | $(0)^2 = 0$ |
| $X_6 = 6$ | $6 - 6 = 0$ | $(0)^2 = 0$ |
| $X_7 = 6$ | $6 - 6 = 0$ | $(0)^2 = 0$ |
| $X_8 = 7$ | $7 - 6 = 1$ | $(1)^2 = 1$ |
| $X_9 = 7$ | $7 - 6 = 1$ | $(1)^2 = 1$ |
| $X_{10} = 8$ | $8 - 6 = 2$ | $(2)^2 = 4$ |
| $X_{11} = 10$ | $10 - 6 = 4$ | $(4)^2 = 16$ |
| $\Sigma X = 66$ years | $\Sigma(X - \bar{X}) = 0$ | $\Sigma(X - \bar{X})^2 = 44$ |
| $\bar{X} = \Sigma X \div n = 6$ years; $n = 11$; $n - 1 = 10$ | | |

Now that we have calculated the mean of the X values ($\bar{X} = 6$ years) we can proceed with the calculation of the sample variance and the sample standard deviation.

First we must find the deviation of each child's age from the mean (average) age. Notice that the second column of numbers in Table 5.6 represents the subtraction of the mean age from each child's age. The resultant value in each case is the number of years the child's age deviates from the mean of the group. For example, the first child ($X_1$) is two years old. The deviation between this child's age and the mean is $-4$ years. The eleventh child ($X_{11}$) is 10 years old. The deviation between this child's age and the mean is four years. Notice that the sum of these 11 deviations from the mean is equal to zero, $\Sigma(X - \bar{X}) = 0$.

Second, we must square each of the deviations. The deviations are squared in the third column of Table 5.6.

Third, we sum all of these squared deviations from the mean. The sum of the squared deviations from the mean, $\Sigma(X - \bar{X})^2$, is equal to 44. The sum of the squared deviations from the mean appears at the bottom of the third column.

Fourth, the sample variance is the sum of the squared deviations from the mean divided by $n - 1$, the number of children's ages. The number of children's ages (n) minus one ($n - 1$) represents what is known as degrees of freedom (df). We will discuss degrees of freedom (df) in subsequent chapters. The symbol for the sample variance is $s^2$. Therefore,

$$s^2 = \frac{\Sigma(X - \overline{X})^2}{n - 1},$$

$$= \frac{44 \text{ years squared}}{11 - 1},$$

$$s^2 = 4.4 \text{ years squared.}$$

The sample standard deviation, s, is the square root of the sample variance. Therefore,

$$s \text{ (the sample standard deviation)} = \sqrt{s^2} \text{ (the square root of the sample variance)},$$

$$s = \sqrt{4.4 \text{ years squared}},$$

$$s = 2.10 \text{ years.}$$

The formula for the calculation of the standard deviation of a sample using this mean-deviation method is,

$$s = \sqrt{\frac{\Sigma(X - \overline{X})^2}{n - 1}}.$$

This mean-deviation method was used here to demonstrate that the sample standard deviation represents actual deviations of sample values from the sample's mean value. This mean-deviation formula can be converted mathematically to a computational formula, which is much easier to use. This computational formula is,

$$s = \sqrt{\frac{\Sigma X^2 - \left[\frac{(\Sigma X)^2}{n}\right]}{n - 1}}.$$

Using the same data presented in Tables 5.5 and 5.6 we can demonstrate the procedures for calculating s, the sample standard deviation, using the computational formula.

Table 5.7 demonstrates the procedures for the calculation of s using the computational formula. Note that the calculated values of s are the same regardless of the method used. The computational formula is much easier to use, especially when the mean value is a fraction, which is often the case.

The symbols $\overline{X}$, $s^2$, and s represent the sample mean, variance, and standard deviation, respectively. The symbols $\mu$, $\sigma^2$, and $\sigma$ represent the population mean, variance, and standard deviation, respectively. The computational formulas for these sample statistics are presented with the computational formulas for the respective population parameters in Table 5.8.

The population standard deviation, sigma ($\sigma$), may be calculated using the same data that was presented in Table 5.7. In Table 5.9 the computational formula for the calculation of the population standard deviation is demon-

**TABLE 5.7** Procedures for the Calculation of s, the Sample Standard Deviation, Using the Computational Formula

| Child's Age (X) in Years | $X^2$ | Computation Formula |
|---|---|---|
| 2 | 4 | $s = \sqrt{\dfrac{\Sigma X^2 - \left[\dfrac{(\Sigma X)^2}{n}\right]}{n-1}}$, |
| 4 | 16 | |
| 5 | 25 | |
| 5 | 25 | |
| 6 | 36 | $= \sqrt{\dfrac{440 - \left[\dfrac{(66)^2}{11}\right]}{11-1}}$ |
| 6 | 36 | |
| 6 | 36 | |
| 7 | 49 | |
| 7 | 49 | $= \sqrt{\dfrac{440 - 396}{10}}$, |
| 8 | 64 | |
| 10 | 100 | |
| $\Sigma X = 66$ | $\Sigma X^2 = 440$, where n = 11 | $= \sqrt{4.4}$ |
| | | s = 2.10 years |

**TABLE 5.8** The Computational Formulas for the Calculation of the Means, Variances, and Standard Deviations for Both Samples and Populations

| | Sample Statistic | Population Parameter |
|---|---|---|
| **Mean** | $\overline{X} = \dfrac{\Sigma X}{n}$, | $\mu = \dfrac{\Sigma X}{n}$ |
| **Variance** | $s^2 = \dfrac{\Sigma X^2 - \left[\dfrac{(\Sigma X)^2}{n}\right]}{n-1}$, | $\sigma^2 = \dfrac{\Sigma X^2 - \left[\dfrac{(\Sigma X)^2}{n}\right]}{n}$ |
| **Standard Deviation** | $s = \sqrt{\dfrac{\Sigma X^2 - \left[\dfrac{(\Sigma X)^2}{n}\right]}{n-1}}$, | $\sigma = \dfrac{\sqrt{\Sigma X^2 - \left[\dfrac{(\Sigma X)^2}{n}\right]}}{n}$ |

strated. However, we must assume that these data, the ages of the children with chicken pox, no longer represent a sample. Instead, we assume these 11 ages represent the finite population of ages of *all* chicken pox sufferers in a small community.

The infinite nature of populations makes it impossible or impractical to actually calculate population parameters such as the population mean ($\mu$) and the population standard deviation ($\sigma$). Therefore, we use sample statistics

**TABLE 5.9** Procedures for the Calculation of $\sigma$, the Population Standard Deviation, Using the Computational Formula

| Child's Age (X) in Years | $X^2$ | Computation Formula |
|---|---|---|
| 2 | 4 | $\sigma = \sqrt{\dfrac{\Sigma X^2 - \left[\dfrac{(\Sigma X)^2}{n}\right]}{n}}$, |
| 4 | 16 | |
| 5 | 25 | |
| 5 | 25 | |
| 6 | 36 | $= \sqrt{\dfrac{440 - \left[\dfrac{(66)^2}{11}\right]}{11}}$ |
| 6 | 36 | |
| 6 | 36 | |
| 7 | 49 | |
| 7 | 49 | $= \sqrt{\dfrac{440 - 396}{11}}$, |
| 8 | 64 | |
| 10 | 100 | |
| $\Sigma X = 66$ | $\Sigma X^2 = 440$, where $n = 11$ | $= \sqrt{4.0}$ |
| | | $\sigma = 2$ years |

($\overline{X}$ and s) to describe sample distributions obtained from populations. These sample statistics also serve to estimate unknown population parameters. In the calculation of the sample standard deviation and variance, $n - 1$ instead of n is used in the denominator to provide a better estimate of the respective population parameters.

When we attempt to describe the distribution of sample values using statements of central tendency and dispersion, the range and interquartile range are most often reported in conjunction with the median value. The standard deviation is always paired with the mean when describing symmetrical, interval-level data.

## SUMMARY

We rely on data description to help us become more informed. Unfortunately, the methods used to describe data fairly and responsibly may also be used to misinform. When reading descriptions of data in professional journals, always cautiously investigate the methods of description. If you discover that another method of description, other than the method used, results in a different interpretation of the data, then you must seriously question the methods and motives of the author. When using descriptive techniques to promote understanding of sample data, the description should summarize, simplify, and organize the data in a way that enhances the data's interpretation.

## FOR REVIEW

1. Discrete data are expressed as whole units, such as the number of patients on a ward, the number of children in a family, or the number of inoculations given.
2. Continuous data are reasonably expressed as fractional units such as miles driven, length of a board, or ages of children.
3. A ratio is equal to the number of cases experiencing a particular event during a particular time period divided by the number of cases experiencing a different event during the same time period.
4. A proportion is equal to the number of cases experiencing a particular event during a given time period divided by the number of cases at risk of experiencing the particular event during the same time period.
5. A rate is equal to the number of cases experiencing a particular event during a certain time period divided by the total number of cases at risk of experiencing the particular event, at a particular time, during that certain time period.
6. A frequency table summarizes the actual counts of cases according to some known interval.
7. A relative frequency table summarizes the fraction of the actual counts of cases divided by the total number of cases in the group according to some known interval.
8. The histogram is a graph in which columns are created proportional to the number of cases or relative cases (vertical axis) and compared with known continuous intervals (horizontal axis).
9. The frequency polygon is a line graph of the actual counts of cases (vertical axis) compared with known continuous intervals (horizontal axis). Relative values may also appear as line graphs.
10. The bar graph is a graph in which columns are created proportional to the number of cases or relative cases (vertical axis) and are compared with discrete categories (horizontal axis).
11. A positively skewed distribution is one in which most of the cases occur toward the negative or left end of the distribution with the tail of the distribution tapering off toward the positive or right end of the distribution.
12. A negatively skewed distribution is one in which most of the cases occur toward the positive or right end of the distribution with the tail of the distribution tapering off toward the negative or left end of the distribution.
13. A mesokurtic distribution is similar to the symmetrical, bell-shaped, normal distribution.
14. A leptokurtic distribution is a distribution of homogeneous values that is more peaked than the normal distribution.
15. A platykurtic distribution is a distribution of heterogeneous values that is more flattened than the normal distribution.
16. The mean is the average, the location in the distribution of values at which the deviations above it and the deviations below it are equal.

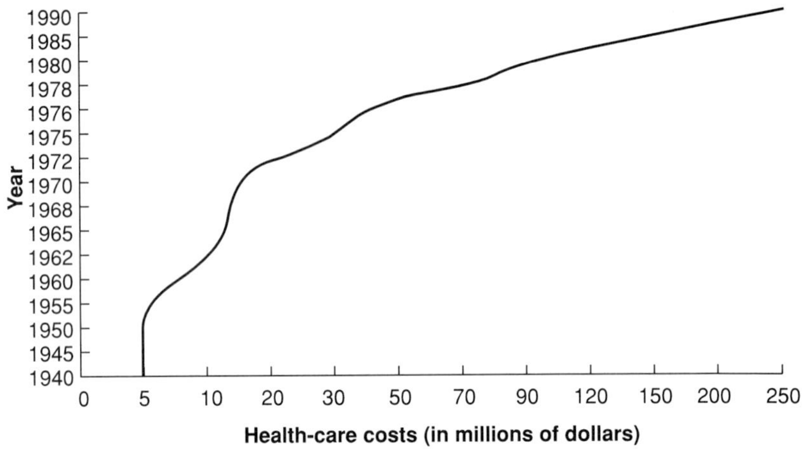


**FIGURE 5.16**
Progression of health-care costs (1940–1990).

17. The median is the middle value in a distribution of ordered values. The median is the value above which there are the same number of values as below.
18. The mode is the value that occurs in a distribution of values most frequently. The mode may not exist and, if it does, may not be unique.
19. Symmetrical distributions have means and medians that are equal.
20. The range is a statement of dispersion found by subtracting the lowest value in the distribution from the highest. It is the distance from one end of a distribution to the other.
21. The interquartile range is the range of the values that comprise the middle 50% of the values in a distribution. It is not sensitive to asymmetry.
22. The average deviation is the amount of absolute (positive and negative) deviation from the mean value exhibited by a distribution of values.
23. The variance of a distribution of sample values is the sum of the squared deviations from the mean value divided by the number of values in the distribution minus one.
24. The standard deviation is the square root of the variance.
25. Descriptive statistics include those techniques used to summarize and present sample data.

### Chapter Knowledge and Comprehension Application

1. Observe the line graph above (Figure 5.16). Comment on the graph's construction. Is a reasonable interpretation of the relationship between health-care costs and time in years possible? If so, why? If not, why not?
2. The nurses at a local hospital have entered into salary negotiations with the hospital administration's representatives. Both sides have access to the information in Figure 5.17. The national distribution of nursing salaries at the time of the negotiations was symmetrical with a mean of $35,000 and a standard deviation of $3,000. Which statements of central

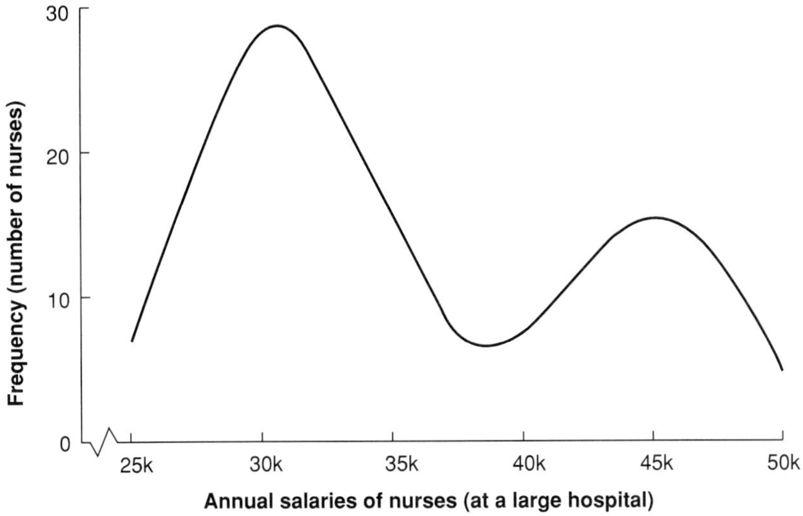


**FIGURE 5.17**
The distribution of salaries earned by nurses in a large hospital.

tendency and dispersion will be reported by the nurses to make their case for salary increases? How might the hospital's administration choose to describe the nurse's salaries in order to hold the line on costs? How would you describe the distribution of salaries in Figure 5.17?

3. When measurements are accurate they are both precise and unbiased. **Unbiasedness** is indicated if a set of measurements are equal to a standard value. The mean of a set of measurements is often compared to the standardized known value to indicate unbiasedness. **Precision** refers to the degree of dispersion of a set of repeated measurements and is typically measured using the standard deviation. The less the dispersion from the known standard, the more precise the measurements.

   A laboratory facility has decided to purchase a new instrument to be used to measure levels of total cholesterol in the blood. Three popular manufacturers were contacted and each supplied the laboratory facility with a new instrument, recently calibrated, for comparison. Standardized samples of blood containing 210 mg of total cholesterol/100 ml of blood were prepared and used to test the accuracy of the three instruments. Each manufacturer's instrument was used to measure 10 of the standardized samples. The means and standard deviations of the 10 trials were calculated for each instrument. These are presented below:

| **Instrument** | **Mean (mg/100 ml cholesterol)** | **Standard Deviation (mg/100 ml cholesterol)** |
|---|---|---|
| 1 | 209.8 | 10.10 |
| 2 | 208.9 | 10.10 |
| 3 | 209.8 | 17.00 |

Which instrument would you buy if accuracy of measurement was the only consideration?

4. A certain laboratory test was used to help diagnose a disease. We also collected information to help determine if the test was effective. Observe the test results below:

| | **Test Result** | | |
|---|---|---|---|
| **Disease Status** | **Negative** | **Positive** | **Total** |
| Absent | 490 | 100 | 590 |
| Present | 10 | 90 | 100 |
| **Total** | 500 | 190 | 690 |

What is the proportion of those with the disease who were reported positive for the disease by the test? (This is the sensitivity of the test.)

What is the proportion of those without the disease who were reported negative for the disease by the test? (This is the specificity of the test.)

What is the proportion of those who tested positive who *did not* have the disease? (false positives)

What is the proportion of those who tested negative who actually had the disease? (false negatives)

## Answers

1. This graph is an example of all that is bad in graphic data display. The impression given is exactly the opposite of what the data indicates. First, notice that time (years) is on the vertical axis. So when the graph goes straight up (1940–1950), health-care costs were not increasing at all ($5 million). And when the graph is relatively flat (1976–1990) the cost of health care was zooming upward ($50 million to $250 million). The graph gives the impression that is the opposite of the truth. Notice also that the time scale (years) and the cost scale (dollars) do not have units that are consistently equal along either axis. It is, therefore, impossible to compare the rise in health-care costs for any particular time interval.
2. Since we know that the distribution of nurses' salaries nationally is symmetrically about a mean of $35,000 ± $3,000, we can assume that the median for this distribution is also $35,000.

   The nurses would probably report the median and range to describe their salary distribution at the large hospital. The median salary would be about $31,000 with a range of about $26,000. This would support their contention that they are paid less than the median national salary for nurses.

   In an effort to hold down costs, the hospital's administrators would most likely report the mean and the standard deviation. The mean of the salaries of the nurses at the large hospital would be very similar to the average salary for the nation.

   Since the data displayed graphically in Figure 5.17 is so obviously positively skewed, the median and interquartile range would probably be the most appropriate statements of central tendency and dispersion.

3. Instrument number 1 produced the most accurate results. This is the one to buy.

| Instrument | Mean | Standard Deviation | Accuracy |
|---|---|---|---|
| 1 | 209.8 | 10.10 | Unbiased, precise |
| 2 | 208.9 | 10.10 | Biased, precise |
| 3 | 209.8 | 17.00 | Unbiased, not precise |

4. Sensitivity = 90/100 = .90 = 90%
   Specificity = 490/590 = .83 = 83%
   False positives = 100/190 = .53 = 53%
   False negatives = 10/500 = .02 = 2%

CHAPTER

6

# Probability

*"A pinch of probability is worth a pound of perhaps."*

**JAMES THURBER**

None of us can get through a single day without consciously or unconsciously making statements of probability. For example, how do we decide if we should dress for rain? Usually we try to determine the likelihood of rain by looking out the window at the clouds or by listening to a weather report. How do we decide when we should leave home to arrive at work or class on time? We usually try to determine how long the trip will take by considering such factors as the chance of encountering traffic congestion or whether we'll find a suitable parking place.

Almost every decision we've ever made has involved an element of chance. When faced with important decisions, we attempt to reduce our chances of making a bad decision by learning as much as we can about all of the factors that could influence our decision. The more relevant the information we have to consider, the better our chances for making a good decision. For example, when trying to decide whether or not to dress for rain, we would probably listen to the weather report. We would be more likely to decide to wear our rain gear if the weather report predicted an 80% chance of rain as opposed to a 10% chance. Our decision might also be influenced by considering the likelihood of being caught in the rain, if, in fact, it did rain. For example, if we were going to work and could park in a parking garage connected to our office, we might decide against dressing for rain, even if the forecast called for an 80% chance of rain. On the other hand, we may decide rain gear is a necessity if we have planned an outing at the park.

Whether we are considering going on a fishing trip, changing our diets, or having a particular surgical procedure, we are all interested in what will *probably* happen as a result of our decision.

In articles published in professional research journals the authors attempt to make good decisions concerning research findings, just as we attempt to make good decisions in our day-to-day lives. Just as we can never be absolutely sure of our prediction of rain, researchers are never absolutely sure drug A is better than drug B for the treatment of a disease. This is why researchers include qualifying statements of probability when they report their findings.

For example, imagine that researchers have reviewed all the relevant information concerning drug A and drug B and have conducted a sound, controlled clinical trial in which drug A seemed to be better than drug B. When the researchers report their decision, they often report something like, "Based on the results of the study, we conclude that drug A is better than drug B ($P \leq .01$)." This indicates the authors' belief that drug A is better than drug B by using the qualifying probability statement $P \leq .01$. This qualifier is read as "the probability is less than or equal to .01." What the authors mean is they are willing to claim that drug A is better than drug B, with the qualification that a probability of .01 (1 chance out of 100, or a 1% chance) or less exists that the findings may have been the result of chance errors. Another way to interpret this decision is that these researchers can be confident that 99% of the time the observed difference between responses to drug A and responses to drug B will be due to something *other than chance*. Or, given sound experimental design, the difference between the responses to drugs A and B is probably due to non-chance factors.

In most published research, the largest qualifying probability statement reported is $P \leq .05$. This means that the researcher's decision is made with only a 95% level of confidence that the differences observed are due to something other than chance. This leaves a 0.5% probability that the observed differences are due to chance. Smaller qualifying probabilities such as $P \leq .001$ are used when high levels of non-chance confidence are important. In Table 6.1 some qualifying probabilities commonly reported in professional literature have been presented. Beside each commonly reported qualifying probability, the associated probability of chance error and the level of non-chance confidence are also reported.

Notice that of the qualifying probabilities presented, $P \leq .05$ has the largest probability of chance error (5% or less), while $P \leq .001$ has the smallest probability of error (0.1% or less). Notice also that as the probability of chance error decreases, the level of non-chance confidence increases.

It was mentioned earlier that smaller qualifying probabilities are used when high levels of non-chance confidence are reported. For example, if the results of the study indicate that drug A is better than drug B, we might decide

**TABLE 6.1** Common Qualifying Probabilities and Levels of Confidence

| Qualifying Probability | Probability of Chance Error (Chance Differences) | Level of Non-chance Confidence (Non-chance Differences) |
|---|---|---|
| $P \leq .05$ | 5% or less | 95.0% |
| $P \leq .025$ | 2.5% or less | 97.5% |
| $P \leq .01$ | 1% or less | 99.0% |
| $P \leq .001$ | .1% or less | 99.9% |

to recommend the use of drug A over the use of drug B. If we also knew that drug B was the most commonly prescribed drug for the effective treatment of a life-threatening disease, then what percentage of non-chance confidence should we require before recommending drug A over drug B? In other words, if we know that our decision could dramatically affect the health of many people, what chance of error should we accept? To answer the first question, our recommendation of drug A over the effective drug of choice for treatment of a life-threatening disease should be made only if we have a very high level of non-chance confidence, perhaps 99.9 or 99.99%. To answer the second question, it seems clear that we should accept only the smallest probability of errors due to chance differences when making decisions with potentially life-or-death consequences, perhaps $P \leq .001$ or $P \leq .0001$.

Before we continue with our discussion of probability, it is important to note that researchers do not arbitrarily claim to know, with a probability of $P \leq .05$ or $P \leq .01$, that drug A is better than drug B. That is, these qualifying probabilities are determined by inferential statistical tests and the use of various statistical tables of probability. Some of these tests and tables will be discussed later. Even after completion of these tests the researcher only knows, with some confidence, whether chance can be eliminated as a probable cause of the differences.

**Probability,** then, may be defined as the relative frequency with which an event occurs or as the proportion of all possible events that are of a particular type. Probability defined as the relative frequency with which an event occurs may be referred to as experience-based probability. When it is possible to know beforehand all the possible outcomes of events and that each outcome has an equal chance of occurrence, we can base our probability statements on logic.

## PROBABILITY BASED ON EXPERIENCE

The relative frequency with which an event occurs (probability of the event's occurrence) may be found by collecting a sample and organizing the sample data to determine the relative frequency of the event. We have already demonstrated this using the blood alcohol concentration percentage (BAC%) data in Table 5.3.

These data are organized into four BAC% intervals in Table 5.4 and presented as frequencies, relative frequencies, and relative frequency percents. From the relative frequency column we can determine the probability of a subject's BAC% falling in any one of the four listed intervals. For example, based on these 30 cases (Table 5.4), what is the probability of a subject's BAC% falling in the third interval (0.10–0.14)? As you can see from the frequency column, the BAC percents for 6 of the 30 cases are in the 0.10–0.14 interval. The relative frequency for cases demonstrating BAC percents in the 0.10–0.14 interval is $6 \div 30$ or $\frac{6}{30}$. Therefore, the probability of a subject's BAC% falling in the 0.10–0.14 interval (the probability of a specific event) is equal to the number of cases experiencing that event (6) divided by the total number of cases (30) who might have experienced the event. This then is $6 \div 30 = .20$, or a probability of .20.

In many health-related fields (medicine, insurance, occupational safety management, and nursing) we often have no logical scheme to help us determine the probability of events. In that case we must depend upon experience and the determination of relative frequency to help us estimate the probability of these events. In other fields (toxicology, industrial hygiene, chemistry, and engineering) logical models are used to predict outcomes. In these cases it is not necessary to collect a sample, organize the data, and determine relative frequencies. Instead, logic is used to determine the probability of certain events.

## PROBABILITY BASED ON LOGIC

In some situations we are able to determine all the possible outcomes of an event and that each outcome has an equal chance of occurring. In these situations we can use logic to determine the probability of an event without relying only on experience.

For example, consider the coin flip. We could flip a coin many times and record the number of times we get heads. In this way we could, by recording our experience, establish the probability of getting heads as an outcome of a coin flip. However, since it is possible to know all the possible outcomes of a coin flip, and since we know, with a fair coin, that all the outcomes (in this case heads or tails) have an equal chance of occurring, then we can use logic to determine the probability of getting heads.

Since a flip of a coin can result in a total of two possible outcomes (heads and tails), and since heads is one of these outcomes, the logical probability of heads is 1/2 or $P = .50$. It is important to remember that the result of each coin flip occurs independently. That is, two events (in this case coin flips) are independent occurrences. One result does not alter the probability of the other result occurring. Therefore, the probability that the next flip of the coin will be heads is also $P = .50$. Even if the outcomes of five previous flips have been tails, the probability of flipping heads the next time is $P = .50$.

## RULES OF PROBABILITY

There are two basic rules to remember when computing probabilities. The first, the **multiplication rule,** states that the probability of the common occurrence of two or more *independent events* is found by *multiplying* their separate probabilities. Remember, events are independent when the outcome of the first event does not influence nor is influenced by the outcome of the second event. As we have shown, the logical probability of getting heads on the first flip of a coin is P = .50. Likewise, the probability of heads on the second flip is also P = .50. Therefore, the probability of two heads in two flips is .50 × .50 = P = .25.

The second rule is the **addition rule.** This rule states that probability of the occurrence of two or more **mutually exclusive events** is found by *adding* their separate probabilities. Events are mutually exclusive if the occurrence of one event prohibits the occurrence of the other. For example, what is the probability of rolling a die and having a three or a six appear? The logical probability of rolling a three is P = 1/6 or P = .167. Also, the probability of rolling a six is P = 1/6 or P = .167. These events are mutually exclusive. If we roll a three we cannot roll a six on the same roll. Therefore, the probability of rolling a die and having a three *or* a six appear is .167 + .167 = .334.

## SUMMARY

We use statements of probability to help us make decisions. We have seen that we often express statements of probability as percentages. When the weather reporter suggests there is an 80% chance of rain, we know the equivalent probability is .80. Similarly, if an event is certain to occur, that is, will occur at a frequency of 100%, its probability of occurrence is 1.00. If an event cannot occur, its probability is 0.

## FOR REVIEW

1. Probability is the number of times a specific event can occur out of the total possible number of events.
2. Events are independent when the outcome of the first event does not influence nor is influenced by the outcome of the second event.
3. Events are mutually exclusive if the occurrence of one event prohibits the occurrence of the other event.
4. The probability of the occurrence of two or more independent events is found by multiplying their separate probabilities.
5. The probability of the occurrence of two or more mutually exclusive events is found by adding their separate probabilities.

**TABLE 6.2** Frequency, Relative Frequency, and Relative Frequency Percentages for Total Blood Cholesterol Levels (mg per deciliter) for 50 Adult Males

| Total Cholesterol (mg/decl.) | Frequency | Relative Frequency | Relative Frequency Percent |
|---|---|---|---|
| 120–139 | 6 | .12 | 12% |
| 140–139 | 8 | .16 | 16% |
| 160–179 | 10 | .20 | 20% |
| 180–199 | 10 | .20 | 20% |
| 200–219 | 8 | .16 | 16% |
| 220–239 | 6 | .12 | 12% |
| 240–259 | 1 | .02 | 02% |
| 260–279 | 1 | .02 | 02% |
| | 50 | 1.00 | 100% |

**Chapter Knowledge and Comprehension Application**
Observe the information presented in Table 6.2.

1. Based on the information presented in Table 6.2, what is the probability that an adult male will have a total cholesterol level of less than 140?
2. What is the probability that an adult male will have a total cholesterol level of 240 or more?
3. What percentage of cases had cholesterol levels of at least 160 but lower than 200?
4. How many of the cases had cholesterol levels lower than 200?

**Answers**
1. P = .12 (6/50)
2. P = .04 (1/50 + 1/50, addition rule)
3. 40% (10/50 + 10/50)
4. 34 cases (6 + 8 + 10 + 10)

CHAPTER

7

# The Normal Distribution

*"Chance is always powerful. Let your hook be always cast. In the pool where you least expect it will be a fish."*

**OVID**

In Chapter 5, we learned that data distributions may be described in several ways. One method, the graph, was used to describe peaked, flat, skewed, and normal distributions. In this chapter we will concentrate on the characteristics of the normal distribution. The normal distribution is also known as the Gaussian distribution, named for the German mathematician and astronomer Carl Friedrich Gauss (1777–1855), who is customarily given credit for its discovery.

Before we begin our discussion of the characteristics of the normal distribution, it is important to keep two things in mind. First, remember that the **normal distribution** is a frequency distribution with continuous, randomly occurring data plotted on the X axis and frequency (counts) plotted on the Y axis. Second, keep in mind that this normal distribution is actually a theoretical distribution. Many real-world distributions are similar to this ideal, theoretical normal curve described by Gauss. We can, therefore, use the theoretical normal curve to help us generate frequencies, percentages, and probabilities for these real-world situations in much the same way that we generated probabilities based on experience in Chapter 6.

## CHARACTERISTICS OF THE NORMAL DISTRIBUTION

We will discuss four important characteristics of the normal distribution that, *together,* differentiate it from other frequency distributions:

1. In Figure 7.1 we can see the basic shape of a frequency distribution in which data are distributed normally. Frequency graphs of normally distributed data are more or less **bell-shaped.** Notice that the distribution has a center point and two tails.

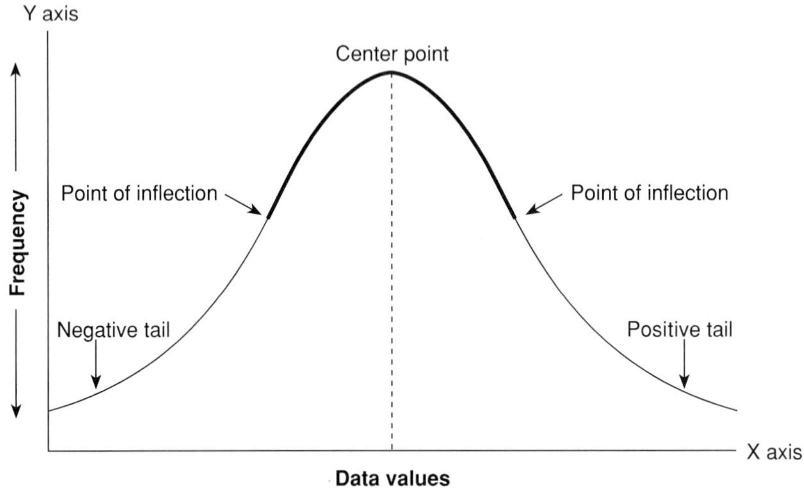


**FIGURE 7.1**
The normal curve: bell-shaped.

The positive tail is the tail of the distribution above and to the right of the center of the distribution, and the negative tail is the tail below and to the left of the center. Notice that the frequency (number of cases) is greatest in the center of the distribution of data values. Also, the frequency of cases decreases as we move farther from the center of the distribution toward either tail. The frequency curve drops off quickly near the center of the distribution to the points of inflection. The frequency curve then falls off less steeply as we move from the points of inflection toward the tails of the distribution. Note also that the tails of the normal distribution never actually touch the X axis. This characteristic makes it theoretically possible for data values to be very large or very small, although the probability such values will occur decreases (with the decrease in frequency) as we move farther from the center of the distribution. Figure 7.1 could serve to represent the shapes of many different real-life distributions such as the heights of all 25-year-old women or the total cholesterol levels of all 40-year-old men. At this time,

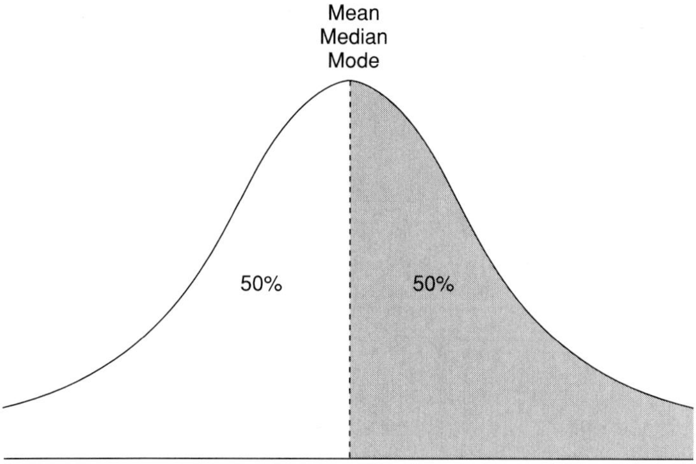


**FIGURE 7.2**
The normal frequency curve, symmetrically distributed about the center point representing the mean, the median, and the mode.

since we are only interested in examining the characteristics of the normal distribution, actual frequencies and data values do not appear on the graph in Figure 7.1. In Figure 7.2 and other figures, the axis indicators, data values, frequency (F), and the vertical axis have been omitted for convenience.

2. The normal distribution is **symmetrical** about its center point. In Figure 7.2 we have shaded the upper half (the half above and to the right of the center point) of the distribution. Notice that the shaded half is the identical mirror image of the unshaded half. Half of the data values fall in the shaded portion and half fall in the unshaded portion. Normal distributions are symmetrical about their center points. However, not all symmetrical distributions are normally distributed. For example, see the non-normal, yet symmetrical distribution in Figure 7.3.
3. Since the normal distribution is symmetrical, the three statements of central tendency (the mean, the median, and the mode) all fall in the *exact same place* in the distribution: the center point or midpoint (see Figure 7.2). In other words, a normally distributed (hence symmetrical) set of data has as its mean a value which is also the most frequently occurring value (the mode) and the value that divides the distribution of data exactly in half (the median). As distributions become asymmetrical, or skewed, the statements of central tendency are no longer identical. With positively skewed data distributions, the mean will move toward the positive tail (to the right) of the distribution and will be greater than the median value (see Figure 7.4). With negatively skewed data distributions, the mean will move toward the negative tail (to the left) of the distribution and will be less than the median (see Figure 7.5).

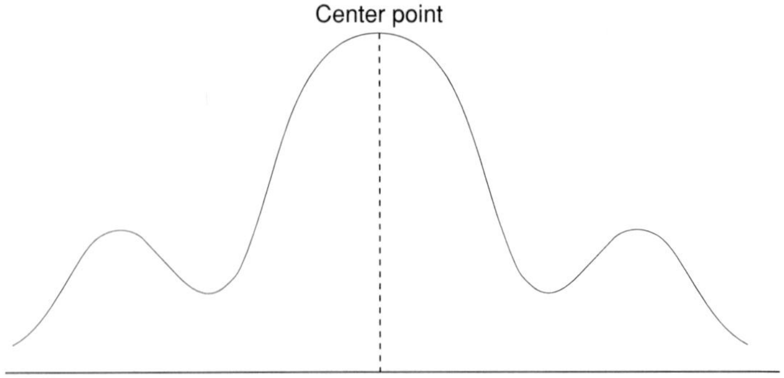


**FIGURE 7.3**
A non-normal, symmetrical frequency distribution.

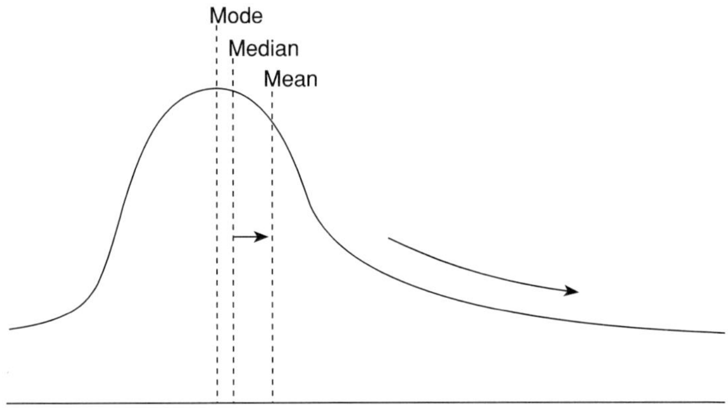


**FIGURE 7.4**
A positively skewed frequency distribution showing the relationship between the statements of central tendency. The mean moves away from the median in the direction of the skew.

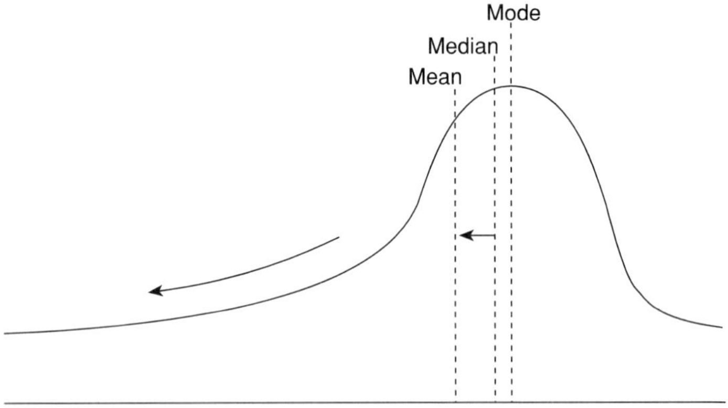


**FIGURE 7.5**
A negatively skewed frequency distribution showing the relationship between the statements of central tendency.

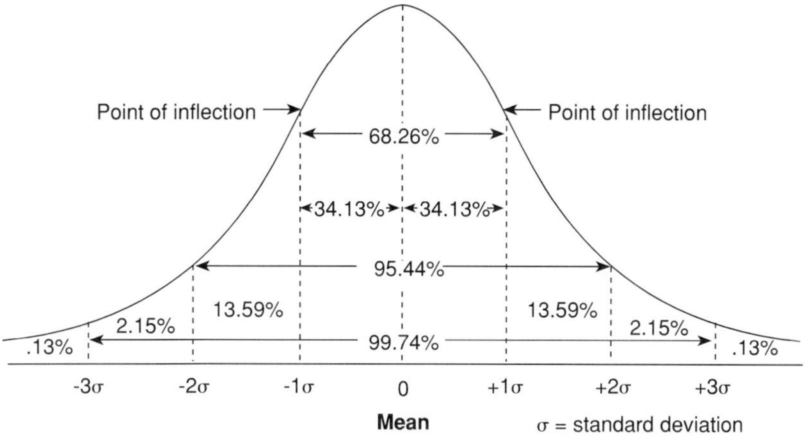


**FIGURE 7.6**
The standard normal curve.

4. The normal distribution may be divided into **constant standard deviation** units ($\sigma$ units). This division results in exact percentages, as shown graphically in Figure 7.6. One standard deviation ($\sigma$) from the mean of the normal distribution occurs at each point of inflection. That is, one standard deviation above the mean ($+1\sigma$) is found by drawing a line perpendicular to the X axis (broken line) from the point of inflection above (to the right of) the mean to the X axis. Notice that 34.13% of the area under the frequency curve is found between the mean and $+1\sigma$. Since the normal distribution is symmetrical, a line drawn perpendicularly from the point of inflection below the mean to the X axis will also include 34.13% of the area under the curve, or 34.13% of the cases. We can also see in Figure 7.6 that between $-1\sigma$ and $+1\sigma$ ($\pm\sigma$) we have a total of 68.26% of the area under the normal curve, or 68.26% of the cases.

Now that we have determined the distance we must travel from the mean in either direction to reach either $-1\sigma$ or $+1\sigma$, we can travel out *twice* that far to either $-2\sigma$ or $+2\sigma$. Also in Figure 7.6, 95.44% of the area under the normal curve is found between $-2\sigma$ and $+2\sigma$ ($\pm 2\sigma$). Notice that only 13.59% of the area under the curve is between $+1\sigma$ and $+2\sigma$, and similarly between $-1\sigma$ and $-2\sigma$.

The area between $+2\sigma$ and $+3\sigma$ includes only about 2.15% of the total area under the normal curve. Because of the symmetrical nature of the normal curve, the area between $-2\sigma$ and $-3\sigma$ also includes about 2.15% of the total area.

Almost all of the area under the normal curve (99.74%) is included between +3 and −3 standard deviations from the mean ($\mu \pm 3\sigma$). The percentage of the normal distribution outside $\pm 3\sigma$ is about 0.26% (100% − 99.74% = 0.26%). This fraction of the area under the curve *beyond*

±*3 standard deviations* is split so that we have 0.13% in each tail of the curve (above $+3\sigma$ and below $-3\sigma$).

We can see that about 34% of the frequency distribution, or about 34% of the cases included in this theoretically normal distribution are found between the average data value (the mean) of the population of values and the point represented by the mean $+1\sigma$ ($\mu + \sigma$).

We can use what we have learned about the normal distribution to find probabilities. The probability (P) that a data value will fall between a value equal to the mean and a value equal to the mean $+1\sigma$ is about 0.34, or P = 0.34. Also, we know the probability of a data value falling between $-1\sigma$ and $-2\sigma$ is equal to about 0.14 (0.1359).

These examples help us understand that the total area under the normal curve is equal to 100%. Thus, 50% of the area under the curve is found above and 50% of the area under the curve is found below the center point (median) or the average (mean). Also, we can see that the probability of a particular data value falling somewhere, anywhere, under the curve is equal to 1.00 or P = 1.00. Likewise, the probability of a particular data value falling anywhere above or anywhere below the mean is equal to 0.50.

## STANDARD SCORES

Many of the sets of raw data we encounter are approximately normally distributed. Because of this, we can use the characteristics of the theoretical normal distribution to help us determine probabilities of occurrence and percentages of area under the normal curve for particular raw scores. To accomplish this, we simply change the normally distributed raw data values of interest to standardized scores or values. These standard scores or values are known as z scores or z values. We can then use these z scores in conjunction with the standard normal curve to identify the respective probabilities and/or percentages.

For example, approximately normally distributed raw data such as heights, birth weights, total serum cholesterol levels, and scholastic aptitude scores can be transformed to z scores. Using the same process we can change each of the specific types of raw data to a standard score. The standard scores (z scores) can then be compared with the standard normal curve where the mean = 0 and a standard deviation = 1.00. Before we examine this transformation process we must first identify the symbols used.

The standardized score that is usually called an z score or z value is defined as:

$$z = \frac{X - \mu}{\sigma}, \text{ where}$$

X = the raw data score or value from the normally distributed population,

$\mu$ = the mean of the population of normally distributed values, and

$\sigma$ = the standard deviation of the population of normally distributed values.

So, to transform a specific X value to a z score we first subtract the population's average ($\mu$) from the X value, then divide the difference by the population's standard deviation ($\sigma$). This results in the creation of a z score with a standard normal distribution where mu ($\mu$) is equal to 0 and sigma ($\sigma$) is equal to 1.00.

Let us assume that the total serum cholesterol levels of a given population are normally distributed with a mean ($\mu$) = 200 mg/dl (milligrams per deciliter), and a standard deviation ($\sigma$) = 40 mg/dl. In Figure 7.7 you can see that approximately 68% of this population ($\mu \pm 1\sigma$) has total serum cholesterol levels between 240 ($\mu + \sigma$ or 200 + 40) mg/dl and 160 ($\mu - \sigma$ or 200 − 40) mg/dl.

We can also see that over 99% of this population ($\mu \pm 3\sigma$) has total serum cholesterol levels between 320 (200 + 40 + 40 + 40) mg/dl and 80 (200 − 40 − 40 − 40) mg/dl.

Using the z transformation formula can help us recognize how the raw scores in Figure 7.7 are converted to their respective z scores.

First, transform the raw score 240 mg/dl to a z score. Using

$$z = \frac{X - \mu}{\sigma}, \text{ where}$$

$$X = 240 \text{ mg/dl},$$

$$\mu = 200 \text{ mg/dl, and}$$

$$\sigma = 40 \text{ mg/dl, then}$$

$$z = \frac{240 \text{ mg/dl} - 200 \text{ mg/dl}}{40 \text{ mg/dl}}$$

$$= \frac{40 \text{ mg/dl}}{40 \text{ mg/dl}},$$

$$z = 1.00$$

Second, transform the raw score 80 mg/dl (Figure 7.7) to a z score. Using

$$z = \frac{X - \mu}{\sigma}, \text{ where}$$

$$X = 80 \text{ mg/dl},$$

$$\mu = 200 \text{ mg/dl, and}$$

$$\sigma = 40 \text{ mg/dl, then}$$

$$z = \frac{80 \text{ mg/dl} - 200 \text{ mg/dl}}{40 \text{ mg/dl}}$$

$$= \frac{-120 \text{ mg/dl}}{40 \text{ mg/dl}},$$

$$z = -3.00$$

From Figure 7.7 we can determine that 84% of the total serum cholesterol levels fall at or below a value of 240. Since we know that the mean,

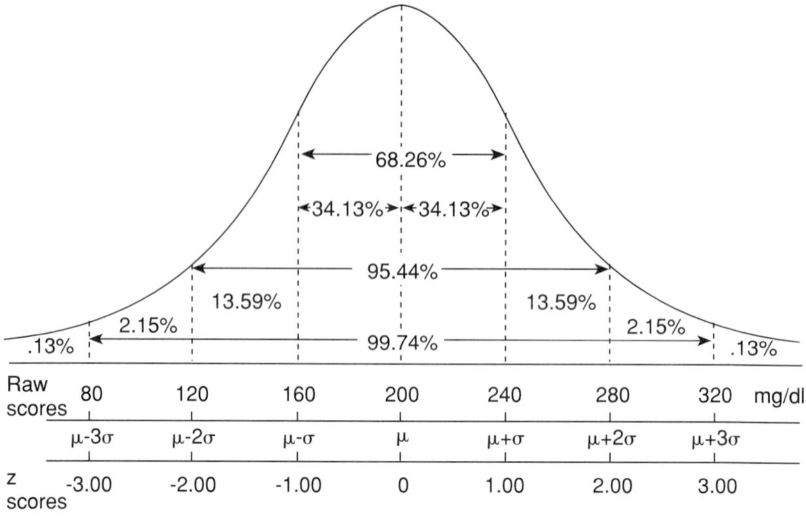


**FIGURE 7.7**
Single standard deviation (σ) divisions of a normally distributed population of total serum cholesterol values and corresponding z scores where $\mu = 200$ mg/dl and $\sigma = 40$ mg/dl.

median, and mode all fall at the center of a normal distribution, and that the center point divides the distribution into two halves, then half of the scores fall below and half fall above the center point. Therefore, 50% of the total serum cholesterol levels fall below the center or mean value, 200 mg/dl (the population average, $\mu$). We also know that approximately 34% of the cholesterol values are found between the mean (200 mg/dl) and the mean $+\sigma$ (240 mg/dl). Likewise, about 34% of the cholesterol values are found between the mean and the mean $-\sigma$ (160 mg/dl). Therefore, the percentage of total serum cholesterol values that fall at or below a value of 240 is 34% + 50% = 84%.

What is the probability that a single randomly selected total serum cholesterol value will be equal to or less than 240? We know that the probability of a randomly selected cholesterol value from this population of values falling below 200 mg/dl is 1/2 or $P = 0.50$. We also know that the probability of a randomly selected value falling between 200 mg/dl and 240 mg/dl is equal to approximately 0.34. We, therefore, have a probability of 0.84 (.50 + .34) that a randomly selected total serum cholesterol value will be equal to or less than 240 mg/dl.

What percentage of the total serum cholesterol values represented in Figure 7.7 are equal to or greater than 160 mg/dl? From this figure we can see that the percentage of the total cases under the curve that are at least 160 mg/dl or greater is about 84% of the total when $\mu = 200$ and $\sigma = 40$. In Figure 7.8 this corresponding 84% of the area under the curve has been shaded. Half (50%) of the cases fall above 200 mg/dl and 34% of the cases fall between 160 and 200 mg/dl for a total of 84%.

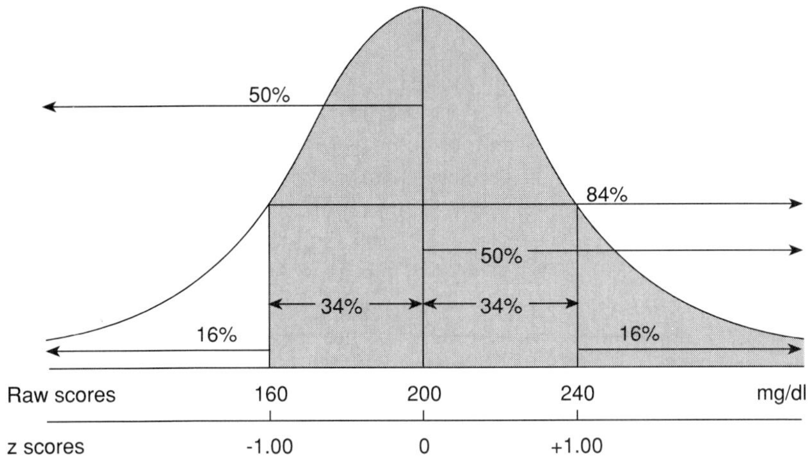


**FIGURE 7.8**
The area under the normal curve corresponding to a total serum cholesterol value of 160 mg/dl or greater when $\mu = 200$ mg/dl and $\sigma = 40$ mg/dl.

What percentage of the cholesterol values represented in Figure 7.8 are less than 160 mg/dl? We can determine that about 16% of the total number of cases under the curve have cholesterol values of less than 160 mg/dl. We can also see that half of the cases (50%) fall below the mean (200 mg/dl), and about 34% of those have cholesterol values of 160 or greater. This leaves only about 16% of the values (50% − 34%) that are less than 160 mg/dl.

From Figure 7.8 we can also determine that about 16% of the cholesterol values are greater than 240 mg/dl. From Figure 7.7 we can determine that the probability of a randomly selected value from this population falling between 240 mg/100 ml and 280 mg/dl, is about 0.14 (0.1359). We can change this probability to a percentage ($0.14 \times 100 = 14\%$) and say that about 14% (13.59%) of the population had cholesterol levels between 240 mg/dl and 280 mg/dl.

We can also use Figure 7.7 to determine standard z scores. For example, what z score corresponds to a raw score of 320 mg/dl? Using

$$z = \frac{320 \text{ mg/dl} - 200 \text{ mg/dl}}{40 \text{ mg/dl}}$$

we can calculate the corresponding z value as 3.00 ($\mu + 3\sigma$ or $200 + 40 + 40 + 40 = 320$ mg/dl). Similarly, for a raw score of 120 mg/dl the corresponding z score is −2.00 ($\mu - 2\sigma$).

Notice that in each of the preceding examples we have found percentages, probabilities, or z scores that corresponded to single, *whole,* standard deviation divisions of the normal distribution. How do we determine the corresponding probabilities, percentages, or z scores associated with *fractional* standard deviation divisions of the normal distribution? Table A in the Appendix can help us determine the areas under the normal curve associated with z scores in increments of 1/100th.

Along its horizontal axis Appendix Table A lists each 1/10th of a z score. These z score fractions begin at 0 and go to 3.0. The vertical axis contains each 1/100th of a z score increment from 0 to 0.09. Notice also that the upper, or positive half of the curve has been superimposed over the chart to help us remember that the numbers within the chart represent proportions of the area under the curve between the z value of interest and $z = 0$, which is the mean ($\mu$) of the theoretical z distribution. The proportions presented in Table A can be converted to percentages by multiplying the proportion of interest by 100. Thus, the decimal point is moved two places to the right and the proportion .3413, for example, becomes $.3413 \times 100 = 34.13\%$.

Also, note that the largest proportion presented in Table A is .4987. This proportion is associated with a z score of 3.00. That is, a z score of 3.0 from the horizontal axis plus 0.00 z from the vertical axis $(3.0 + 0.00)$ equals 3.00. This means that if we move from the theoretical mean of the normal distribution $(z = 0)$ to the right until we have reached 3.00 z, we have included .4987 or 49.87% of the area under the curve. We learned previously that the theoretical normal curve never actually reaches the horizontal axis. It is therefore possible that the curve extends out to perhaps four, five, or six standard deviations (z scores) from the mean. We also know, in theory, that half of the area under the curve may be found above the mean $(z = 0)$ and half below. In Table A, for convenience, the largest z value included is 3.00, at which point we have included 49.87% of the area under the curve.

By definition, the normal curve is symmetrical; therefore, the lower left or negative half of the curve is an exact mirror image of the right upper or positive half. To save space, only the upper or positive half of the curve has been presented in Table A.

What proportion of the area under the normal curve is found between $z = 0$ and $z = -1.00$? From Table A we can see that between $z = 0$ and $z = 1.00$ we have a proportion of .3413. Since the curve is symmetrical, the proportion of the area under the curve between $z = 0$ and $z = -1.00$ is equal to the area between $z = 0$ and $z = +1.00$, or .3413; this equals about 34% (see Figure 7.9a).

What proportion of the area under the curve is found *between* the mean of the normal curve and a z score of $\pm 0.50$? In Table A we can find the z score of 0.50 on the horizontal axis. The corresponding proportion of the area under the curve between $z = 0$ and $z = 0.50$ (1/2 of 1.00 z) is .1915. Because the curve is symmetrical, between $z = 0$ and $z = -0.50$ we also have the proportion .1915, for a total of .3830 (Figure 7.9b).

The percentage of the area under the curve found *between* $z = 0$ and $z = -0.50$ is 19.15% $(.1915 \times 100 = 19.15\%)$. Similarly, the percentage of the area under the curve between $z = \pm 0.50$ is equal to $19.15\% + 19.15\% = 38.30\%$.

We can also find the proportion of the area under the curve *outside* of $z = \pm 0.50$. Since we know that $z = \pm 0.50$ includes .3830 of the total, the proportion of the area under the curve above $z = 0.50$ and below $z = -0.50$ can

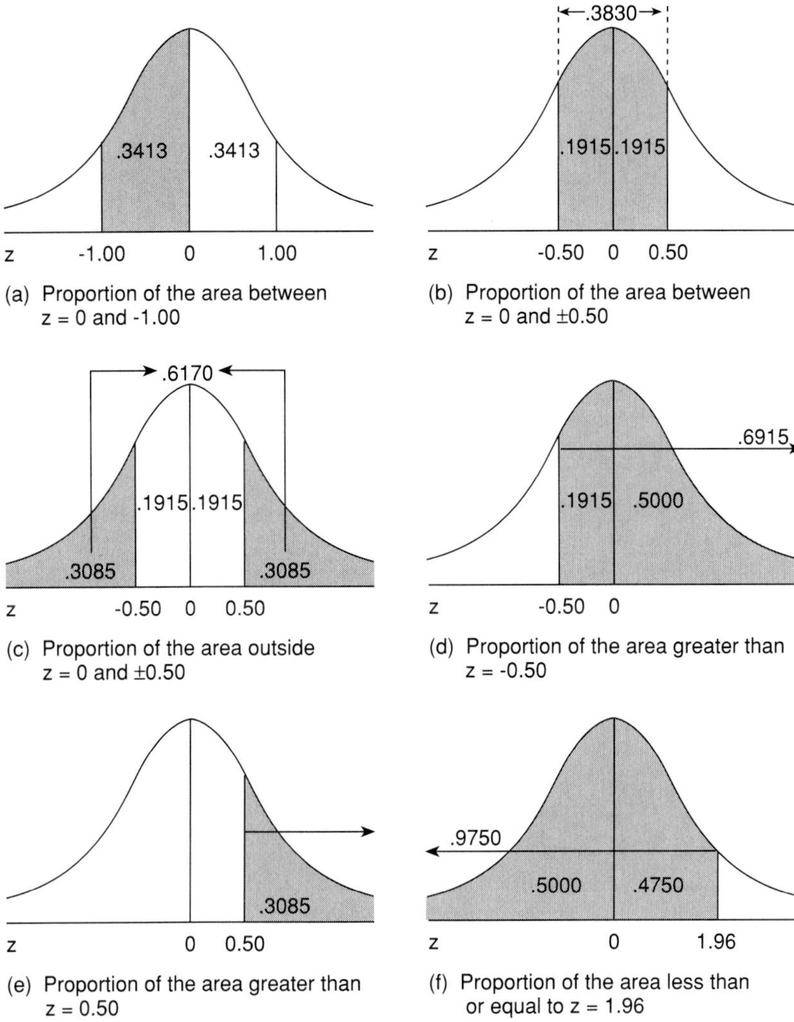


**FIGURE 7.9**
Areas under the normal curve.

be found by subtracting .3830 from 1 (the total area under the curve) to get .6170 (Figure 7.9c).

The proportion of the area under the curve corresponding to a z score *greater than* z = −0.50 is equal to the fraction of the area between z = −0.50 and z = 0 plus the fraction of the area under the curve above z = 0. This is .1915 + .5000 = .6915 (Figure 7.9d).

The proportion of the area under the curve corresponding to a z score *greater than* z = 0.50 is equal to the fraction of the area above z = 0 (.5000) minus the fraction of the area between z = 0 and z = 0.50. This is .5000 − .1915 = .3085 (Figure 7.9e).

What is the proportion of the area under the curve that corresponds with z values *less than or equal to* 1.96? This is found by adding the fraction of the area below $z = 0$ (.5000) to the fraction of the area between $z = 0$ and $z = 1.96$. From Table A we see that the fraction of the area between $z = 0$ and $z = 1.96$ is equal to .4750. Therefore, $.5000 + .4750 = .9750$ (Figure 7.9f).

Suppose we know that the population of total serum cholesterol levels is normally distributed with a mean ($\mu$) of 200 mg/dl and a standard deviation ($\sigma$) of 40 mg/dl.

If your total serum cholesterol level was found to be 180 mg/dl, what percentage of the total serum cholesterol levels in this population were greater than yours?

First, we must transform your raw score to a z score. This transformation will allow us to estimate the percentage of the area under the standard normal curve that corresponds to total serum cholesterol levels greater than your score.

$$\text{Using } z = \frac{X - \mu}{\sigma}, \text{ we have}$$

$$= \frac{180 \text{ mg/dl} - 200 \text{ mg/dl}}{40 \text{ mg/dl}}$$

$$= \frac{-20 \text{ mg/dl}}{40 \text{ mg/dl}},$$

$$z = -0.50$$

In a normally distributed population of total serum cholesterol values having a mean ($\mu$) of 200 mg/dl and a standard deviation ($\sigma$) of 40 mg/dl, your score of 180 mg/dl corresponds to a z score of $-0.50$.

Second, using Table A we can see that .1915 or 19.15% of the area under the curve is found between $z = 0$ and $z = 0.50$. Since the normal distribution is symmetrical, we know this is also true for the area between $z = 0$ and $z = -0.50$ (Figure 7.9b). So we combine this 19.15% with the 50% that falls above the mean ($\mu = 0$) and we have 69.15% (Figure 7.9d). So we can estimate about 69% of the total serum cholesterol values in this example fall above your score, 180 mg/dl.

Given the same normally distributed population of total serum cholesterol values ($\mu = 200$ mg/dl and $\sigma = 40$ mg/dl) we could find the probability of someone from this population having a cholesterol value of equal to or greater than 280 mg/dl.

$$\text{Using } z = \frac{X - \mu}{\sigma}, \text{ we have}$$

$$= \frac{280 \text{ mg/dl} - 200 \text{ mg/dl}}{40 \text{ mg/dl}},$$

$$= \frac{80 \text{ mg/dl}}{40 \text{ mg/dl}},$$

$$z = 2.00$$

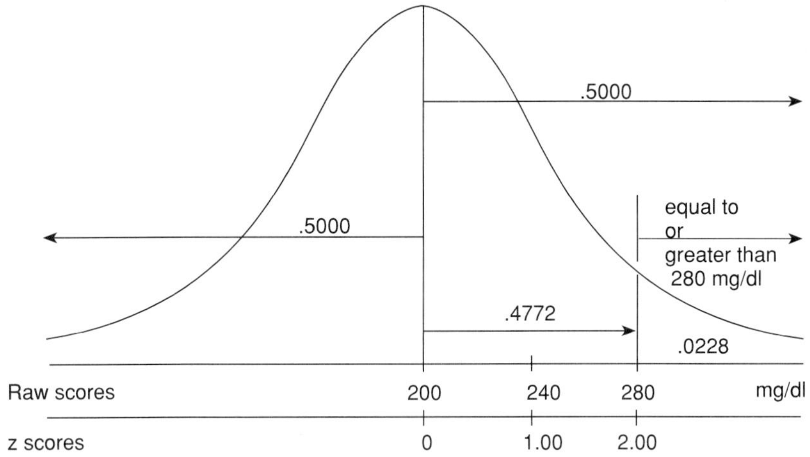


**FIGURE 7.10**
The probability of a randomly selected total serum cholesterol value being equal to or greater than 280 mg/dl given a normally distributed population with $\mu = 200$ mg/dl and $\sigma = 40$ mg/dl.

Using Table A we can see the probability of falling between $z = 0$ and $z = 2.00$ is .4772. To find the probability of having a total serum cholesterol level of 280 mg/dl or greater, we subtract .4772 from .5000, leaving a probability of .0228 (Figure 7.10).

Another way to think of this is that we have a .9772 probability that a cholesterol value will fall at or below 280 mg/dl (the .5000 below $z = 0$, plus the .4772 between $z = 0$ and $z = 2.00$). Therefore, we have a $1.000 - .9772 = .0228$ chance of a total serum cholesterol level of 280 mg/dl or greater occurring in this population.

Using the same normally distributed population of cholesterol values ($\mu = 200$ mg/dl and $\sigma = 40$ mg/dl) let us examine yet another application of the normal distribution. Suppose we were interested in knowing between which two cholesterol values 95% of the values in the population fall (Figure 7.11).

In this case, we have assumed that the two values are equally distant from the mean ($\mu$). Therefore, we divide the interval by two ($95.00 \div 2 = 47.50$) to determine what percentage of the values will occur between the mean and each of the two cholesterol values. From Table A we can see that a z score of 1.96 is associated with the fraction .4750 or 47.50%. This means that 47.50% of the cholesterol values are found between the mean ($\mu = 200$ mg/dl) plus $1.96 \times 40$ mg/dl. Likewise, 47.50% of the cholesterol values are also found between the mean and $-1.96 \times 40$ mg/dl (Figure 7.11).

We know a z score represents the number of standard deviations from the mean. In this case we also know we must go above and below the mean 1.96 standard deviations to include 95% of the cases. Then, to find the actual cholesterol value that occurs at $z = 1.96$, we multiply $1.96 \times \sigma$ (40 mg/dl)

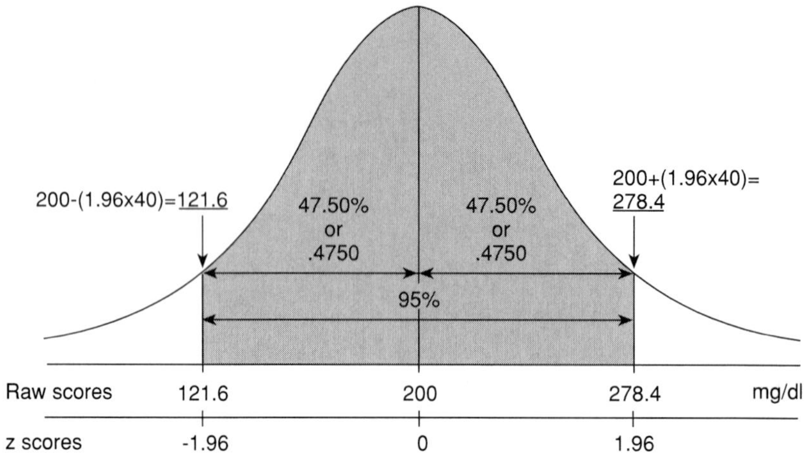


**FIGURE 7.11**
The cholesterol values between which 95% of the values fall when the population is normally distributed with $\mu$ = 200 mg/dl and $\sigma$ = 40 mg/dl.

and add this product to the mean. We have 200 mg/dl + (1.96 × 40 mg/dl) = 200 mg/dl + 78.4 mg/dl = 278.4 mg/dl. Likewise, to find the actual cholesterol value that occurs at z = −1.96 we multiply 1.96 × $\sigma$ (40 mg/dl) and subtract this product from the mean. We then have 200 mg/dl − (1.96 × 40 mg/dl) = 200 mg/dl − 78.4 = 121.6 mg/dl. Therefore, 95% of the cholesterol values in this population fall between approximately 121 mg/dl and 278 mg/dl.

## SUMMARY

The theoretical normal distribution is a frequency distribution that is symmetrical about its mean, median, and mode. The mean, median, and mode of the normal distribution are identical. The continuous data having the greatest frequency are those data values located at and on either side of the center point of the distribution. The continuous data having the least frequency are those data values occurring at the two tails of the distribution.

The theoretical normal distribution can be used to help us answer questions about concrete, real-world distributions. The calculation of z scores can help us determine relative placement of a particular value from a real-world distribution. Also, particular values from unrelated real-world distributions may be compared as z scores in terms of their placement or the theoretical normal distribution.

## FOR REVIEW

1. A z value is the distance between a continuous data value and the population mean divided by the population standard deviation. Therefore, the z value gives the distance between a data value and the mean in units equal to the standard deviation. The z value for the continuous data value X is:

$$z = \frac{X - \mu}{\sigma}$$

where $\mu$ and $\sigma$ equal the population mean and standard deviation, respectively. Note that if $X - \mu$ is equal to $\sigma$ then $z = 1.00$; and if X is equal to $\mu$, then $z = 0$.

**Chapter Knowledge and Comprehension Application**

Given a normally distributed population of diastolic blood pressures where $\mu = 80$ mmHg and $\sigma = 3$ mmHg, find:

1. The z score for a diastolic pressure of 77 mmHg.
2. The percentage of the population with diastolic pressures equal to or less than 77 mmHg.
3. The two diastolic pressures between which approximately 68% of the pressures fall.
4. The diastolic pressure above which approximately 5% of the pressures fall.

**Answers**

1. $z = \frac{77\text{ mmHg} - 80\text{ mmHg}}{3\text{ mmHg}} = \frac{-3\text{ mmHg}}{3\text{ mmHg}} = -1.00$
2. $z = \frac{77\text{ mmHg} - 80\text{ mmHg}}{3\text{ mmHg}} = -1.00$

   From Appendix Table A, the percentage of the area under the curve between $z = 0$ and $z = -1.00$ is 34.13%. The area equal to or less than 77 mmHg ($z = -1.00$) is 50% − 34.13% = 15.87%.
3. 77 mmHg and 83 mmHg (±1.00 standard deviation).
4. Since half of the normal distribution includes 50% of the cases, the score above which 5% of the pressures fall has 45% of the pressures between it and the mean of the population. From Table A, we can see that a z score of 1.65 is associated with .4505 or 45.05% of the cases. Therefore,

$$1.65 = \frac{X - 80\text{ mmHG}}{3\text{ mmHg}}$$

$(1.65 \times 3\text{ mmHg}) + 80\text{ mmHg} = 84.95\text{ mmHg}.$

CHAPTER

# 8

# Estimating the Population Mean Using the Normal Distribution

*"Then there was the man who drowned crossing a stream with an average depth of six inches."*

W. I. E. GATES

In Chapter 7 we concentrated on the normal frequency distribution of a specific population having a mean equal to $\mu$ and a standard deviation of $\sigma$. We considered the population of all cholesterol values and the population of all diastolic pressures. Populations can be infinite in nature, like the population of all rocks on the beaches of the Atlantic Ocean. Populations can also be relatively small, such as the population of all rocks examined by students during Dr. Smith's geology seminar. When defining populations, *all* is the key word.

When your data are taken not from all but from *some* members of the population, you have a sample. A sample is any subset of a population. In Chapter 3 we learned that in most situations the necessary expenditure of time, energy, and money makes a complete and exhaustive review of *all* members of a population prohibitive. Instead, we gather a sample and make inferences about the parameters of the population based on the sample's statistics. Also, we have reviewed several methods for collecting samples from populations and have found that samples that fairly represent all members of the population from which they are collected offer the best chance for making accurate inferences regarding the population's parameters. Random sampling is commonly used to obtain representative samples from a parent population, where that population represents the source of the sample.

## SAMPLING ERROR

Sampling error is always expected. It is not possible for a sample to exactly represent the population. Sampling error does not result from measurement errors or arithmetic errors, although these errors also contribute to inference inaccuracy. Sampling error is the error, attributable to chance, that we make when selecting random samples to represent the population of interest. In other words, **sampling error** is the expected chance difference, variation, or deviation between a random sample and the population.

## THE DISTRIBUTION OF SAMPLE MEANS AND THE CENTRAL LIMIT THEOREM

The distribution of sample means obtained from all possible random samples of the same size taken from the same population has three interesting and useful properties, derived from the **Central Limit Theorem.** First, if we obtain all possible random samples of 25 patients from a known population of all patients, and compute a sample mean ($\overline{X}$) for each of these samples, the shape of this distribution of sample means will be approximately that of the normal curve. As the sizes of the samples taken are increased so as they approach and exceed 30 cases, the shape of the distribution of sample means becomes increasingly similar to the shape of the normal curve.

Second, if we include each sample's mean ($\overline{X}$) in the calculation of a grand mean ($\mu_{\overline{X}}$), the mean of the population of all sample means, the grand mean ($\mu_{\overline{X}}$) will be equal to the mean of the parent population ($\mu$).

Third, the standard deviation of the distribution of means of samples can be determined in the same way we determine the standard deviation of the distribution of individual scores (that is, by substituting the mean scores for the individual scores and using the standard deviation equation). This standard deviation of the sample means serves as a measure of sample variability and is known as the standard error of the mean. The **standard error of the mean (SEM)** indicates the variability of the distribution of means of samples caused by sampling error and by measurement error.

What requirements must be met before we rely on these facts presented as the Central Limit Theorem? The most important requirement is that the size of the sample (n) be large enough. How large a sample is "large enough" depends, for the most part, on the shape of the distribution of the parent population. If the parent population is normally distributed, then n can be small

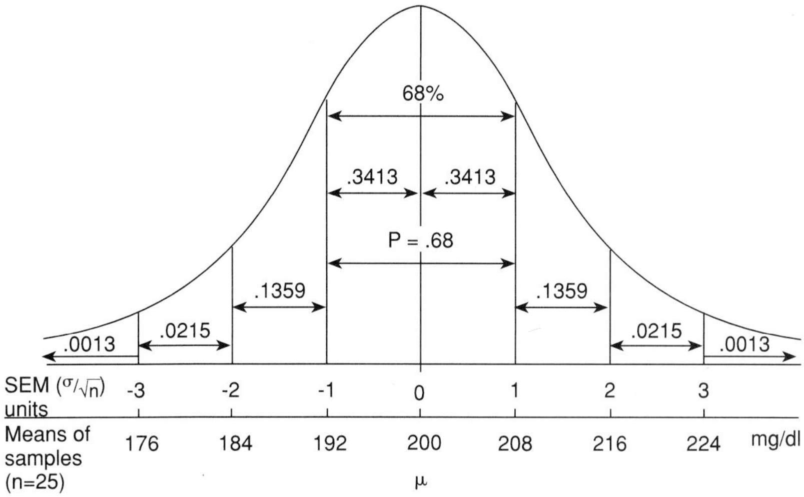


**FIGURE 8.1**
The distribution of sample means (n = 25) from a parent population with a mean (μ) equal to 200 mg/dl total serum cholesterol and a standard deviation (σ) equal to 40 mg/dl. SEM ($\sigma/\sqrt{n}$) = 8 mg/dl total serum cholesterol.

(20 or less). If the population distribution is symmetrical but not bell-shaped, n should be somewhat larger. If the population distribution is clearly non-normal (skewed), n may need to be much larger, perhaps 100 or more. When the shape of the population distribution is unknown, a sample size of 30 or more is usually large enough.

In Figure 8.1 we have a hypothetical distribution of sample means (n = 25). The mean (μ) and the standard deviation (σ) of the parent population are 200 mg/dl and 40 mg/dl, respectively. We can estimate the standard error of the mean from our distribution of sample means as $\sigma/\sqrt{n}$. The SEM is equal to 8 mg/dl. In other words, the standard deviation of the distribution of sample means is 8 mg/dl. Knowing this, and remembering the properties of the distribution of sample means, we can determine the probability of selecting a sample (n = 25) with any particular mean cholesterol value.

From Figure 8.1 we see that the probability of obtaining a sample (n − 25) whose mean is between 200 mg/dl and 208 mg/dl is equal to .3413. Or, approximately 34% of the samples (n = 25) taken from this parent population will have averages between 200 mg/dl and 208 mg/dl. These determinations of probabilities and percentages for the **distribution of sample means** were made exactly in the same manner as were the determinations of probabilities and percentages for the normal distribution of individual scores in Figure 7.6.

## Estimation of the Standard Error of the Mean

We know that the third property of the distribution of repeated random samples taken from a parent population is that the SEM can be determined, theoretically, by substituting the means of the many random samples for the individual scores and using the equation for the calculation of standard deviation. The repeated selection of random samples is often not practical or possible. Instead, we can *estimate* the value of the SEM when less than all of the population's values are known. This estimate of the SEM uses the standard deviation ($\sigma$) of the population of individual or raw scores divided by the square root of the number of scores in the sample. Therefore, we estimate the standard error of the distribution of sample means using

$$SEM = \sigma/\sqrt{n}.$$

In Figure 8.1 we are shown how the SEM is determined to be 8 mg/dl total cholesterol, since the population's standard deviation ($\sigma$) was given as 40 mg/dl total serum cholesterol and the means of the sampling distribution resulted from samples of 25 cases. Therefore, we can now estimate SEM as

$$\begin{aligned} SEM &= \sigma/\sqrt{n}, \\ &= \frac{40 \text{ mg/dl}}{\sqrt{25}}, \\ &= 8 \text{ mg/dl}. \end{aligned}$$

Hence the population mean ($\mu$) plus 1 SEM equals (200 mg/dl + 8 mg/dl) 208 mg/dl. This method of estimating the SEM and the properties of the distribution of sample means makes it possible to infer a population's parameters using sample statistics.

## Importance of the Properties of the Distribution of Sample Means

Because the distribution of sample means approximates the normal distribution, especially when the samples taken are of sufficient size (around 30 cases), we can make probability statements concerning the distribution of sample means ($\overline{X}$) just as we did previously with the distribution of individual scores.

First, let us reconsider the distribution of **individual scores** (X). From a normally distributed population of total serum cholesterol values (mg/dl), where $\mu = 200$ mg/dl and $\sigma = 40$ mg/dl, what is the probability of a single randomly selected cholesterol value (X) being equal to or greater than 192 mg/dl?

$$\begin{aligned} \text{Remember: } X &\geq 192 \text{ mg/dl}, \\ \mu &= 200 \text{ mg/dl}, \\ \sigma &= 40 \text{ mg/dl, and} \\ z &= \frac{X - \mu}{\sigma}. \end{aligned}$$

$$\begin{aligned} \text{We have, } z &= \frac{192 \text{ mg/dl} - 200 \text{ mg/dl}}{40 \text{ mg/dl}}, \\ &= \frac{-8 \text{ mg/dl}}{40 \text{ mg/dl}}, \\ &= -0.20. \end{aligned}$$

Thus, $z = -0.20$ indicates that 192 mg/dl is 2/10ths (8 mg/dl) of one standard deviation (40 mg/dl) below the mean ($\mu$ = 200 mg/dl). From Appendix Table A we see that between $z = 0$ and $z = 0.20$ we have .0793 of the cases. Since the normal curve is symmetrical we know that .0793 of the cases are also found between $z = 0$ and $z = -0.20$. To find the probability of z being equal to or greater than $-0.20$ we add the probability of falling between $z = -0.20$ and $z = 0$ to the probability of z being equal to or greater than 0. Therefore, we have .0793 plus .5000, respectively, for the probability P = .5793 (Figure 8.2).

Now, let us consider the distribution of **sample means** ($\overline{X}$). From a normally distributed population of X values where $\mu$ = 200 mg/dl and $\sigma$ = 40 mg/dl, we can now estimate the probability of selecting a single random sample (n = 25) whose mean ($\overline{X}$) would be 192 mg/dl or greater. To accomplish this we may substitute $\overline{X}$ for X and $\sigma/\sqrt{n}$ for $\sigma$ in our z formula.

$$\text{So, } z = \frac{(X - \mu)}{\sigma} \text{ becomes } z = \frac{(\overline{X} - \mu)}{\sigma/\sqrt{n}}.$$

$$\begin{aligned} \text{Remember: } \overline{X} &\geq 192 \text{ mg/dl}, \\ \mu &= 200 \text{ mg/dl}, \\ \sigma &= 40 \text{ mg/dl}, \\ n &= 25 \text{ mg/dl, and} \\ \sigma/\sqrt{n} &= 8 \text{ mg/dl}. \end{aligned}$$

$$\begin{aligned} \text{Using } z &= \frac{\overline{X} - \mu}{\sigma/\sqrt{n}}, \text{ we have} \\ &= \frac{192 \text{ mg/dl} - 200 \text{ mg/dl}}{40 \text{ mg/dl}/\sqrt{25}} \\ &= \frac{-8 \text{ mg/dl}}{8 \text{ mg/dl}}, \\ &= -1.00. \end{aligned}$$

We estimate that the sample mean ($\overline{X}$) equal to 192 mg/dl is one SEM unit below the expected population mean ($\mu$), resulting in a z score of $-1.00$ (Figure 8.3). From Table A we can see that the estimated probability of a sample average (when n = 25) being 192 mg/dl or greater is equal to the probability of being between $z = -1.00$ and $z = 0$ plus the probability of being greater than $z = 0$. These probabilities are P = .3413 and P = .5000, respectively (.3413 + .5000 = .8413).

Figure 8.3 reveals the probability of P = .8413 of selecting a sample (n = 25) with a mean ($\overline{X}$) total serum cholesterol level of 192 mg/dl or greater.

In the first situation we place 192 mg/dl (an individual score) on the $z = \frac{(X - \mu)}{\sigma}$ distribution. In the second situation we place 192 mg/dl (a mean

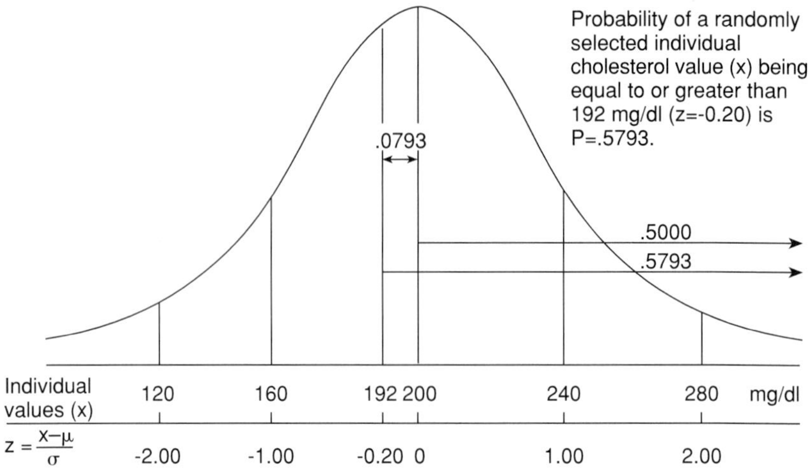


**FIGURE 8.2**
The distribution of mg/dl cholesterol values (X) when $\mu = 200$ mg/dl and $\sigma = 40$ mg/dl of total serum cholesterol showing the probability (P) of a randomly selected individual cholesterol value being equal to or greater than 192 mg/dl.

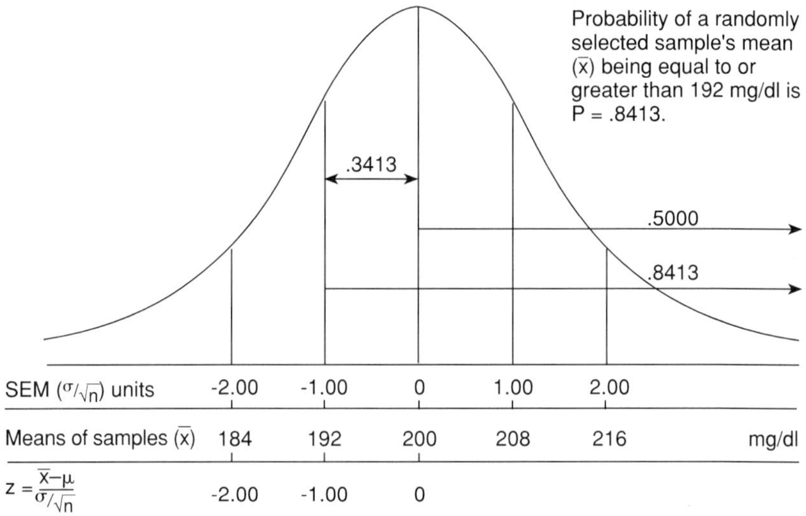


**FIGURE 8.3**
The distribution of sample means ($n = 25$) from a parent population ( $\mu = 200$ mg/dl and $\sigma = 40$ mg/dl) sample's mean ($\overline{X}$) being 192 mg/dl of total serum cholesterol or greater.

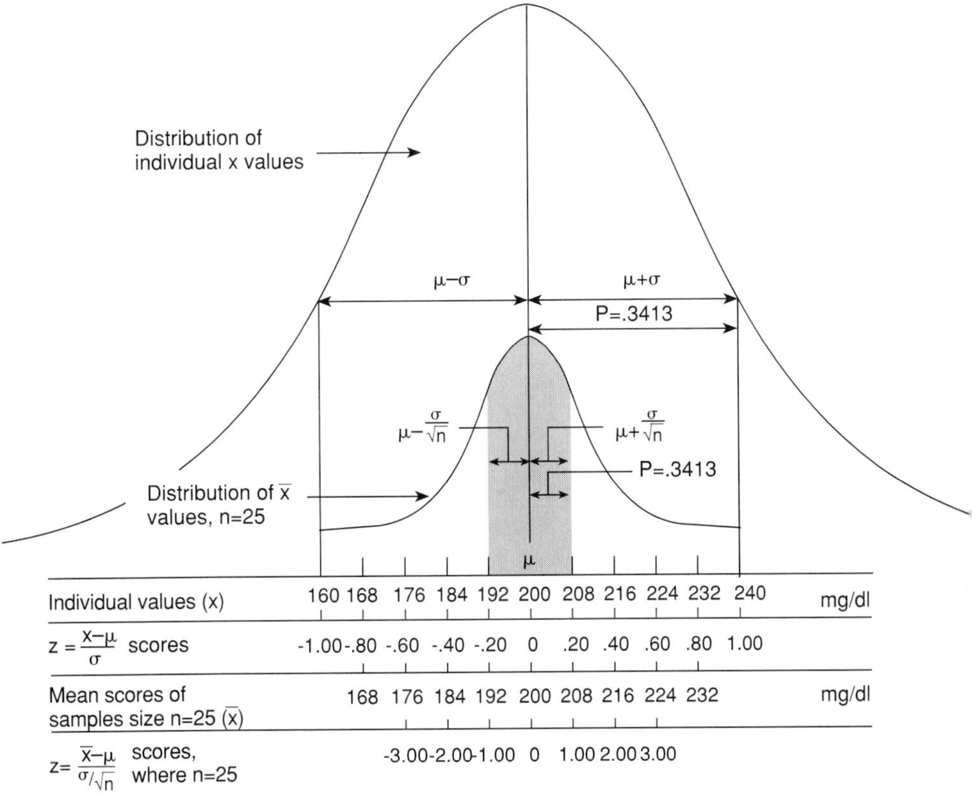


**FIGURE 8.4**
The distribution of individual X values and the distribution of sample means ($\bar{X}$) when $\mu$ = 200 mg/dl and $\sigma$ = 40 mg/dl of total serum cholesterol and the size of the repeated random samples = 25.

of a sample) on the $z = \dfrac{(\bar{X} - \mu)}{(\sigma/\sqrt{n})}$ distribution. As we have seen, the two different z values ($z = -0.20$ and $z = -1.00$) result from dividing by $\sigma$ (40 mg/dl) or $\sigma/\sqrt{n}$ (8 mg/dl), respectively.

In Figure 8.4 we can see that the distribution of individual X values has a standard deviation of 40 mg/dl. One standard deviation below the mean puts the cholesterol value at $z = -1.00$ or 160 mg/dl. One standard deviation above puts the cholesterol value at $z = 1.00$ or 240 mg/dl.

We can also see the distribution of means of samples ($n = 25$). Notice that the estimate of the standard error ($\sigma/\sqrt{n}$) is equal to 8 mg/dl. Therefore, one standard error below the mean (200 mg/dl) puts the average cholesterol value at 192 mg/dl. Likewise, one standard error above the mean puts the average cholesterol value at 208 mg/dl.

**TABLE 8.1** The Effect of Sample Size (n) on the Estimate of Standard Error ($\sigma/\sqrt{n}$) ($\sigma$ = 40 mg/dl, $\mu$ = 200 mg/dl)

| n | $\sqrt{n}$ | $\sigma/\sqrt{n}$ | $\mu \pm \sigma/\sqrt{n}$ |
|---|---|---|---|
| 1 | 1 | 40 mg/dl | 160–240 mg/dl |
| 4 | 2 | 20 mg/dl | 180–220 mg/dl |
| 16 | 4 | 10 mg/dl | 190–210 mg/dl |
| 25 | 5 | 8 mg/dl | 192–208 mg/dl |
| 100 | 10 | 4 mg/dl | 196–204 mg/dl |
| 400 | 20 | 2 mg/dl | 198–202 mg/dl |
| 1600 | 40 | 1 mg/dl | 199–201 mg/dl |

Just as 34.13% of the X values would be expected to fall between 200 mg/dl and 240 mg/dl ($\mu + \sigma$), we expect 34.13% of the $\overline{X}$ *values of samples* (n = 25) to fall between 200 mg/dl and 208 mg/dl ($\mu + \sigma/\sqrt{n}$).

When we are interested in the distribution of means taken from repeated samples of the same size we must adjust $\sigma$ to account for the size of the samples. When we are interested in the distribution of individual X values, since we collect no samples we need not adjust $\sigma$. As the sample size increases, the estimate of standard error becomes smaller and smaller. In Table 8.1, consider the effect of increasing the sample size—from one case per sample to 1,600 cases per sample—on the estimate of standard error when $\sigma$ = 40 mg/dl and $\mu$ = 200 mg/dl.

As the sample size increases from one case per sample (which is essentially the distribution of individual X values, thus $\sigma$ and $\sigma/\sqrt{n}$ = 40 mg/dl) to 1,600 cases per sample, the estimate of standard error is reduced from 40 mg/dl to 1 mg/dl. This means that if we collect repeated samples of four cases per sample, the standard deviation of this distribution of sample means could be estimated as $\sigma/\sqrt{n}$ = 20 mg/dl. We could then estimate that approximately 68% of the sample (n = 4) means would fall between $\mu \pm 1$ SEM. In this case, if $\mu$ = 200 mg/dl and $\sigma/\sqrt{n}$ = 20 mg/dl, then about 68% of the means of samples (n = 4) would fall between 180 mg/dl and 220 mg/dl. If we are able to collect repeated samples of n = 1,600, we could reduce the estimate of standard error to 1 mg/dl. In this case, about 68% of the means of samples where n = 1,600 would fall between 199 mg/dl and 201 mg/dl. Therefore, the bigger the sample size the smaller the estimate of standard error.

We know this intuitively. Which of the following two studies would you select to estimate the total serum cholesterol level of a population of several hundred thousand "first heart attack survivors"? Would you choose to base your inferences on the study having 16 randomly selected subjects or on the study having 1,600 randomly selected subjects? To avoid complication, assume the method and purpose of the two studies were identical except for the size of the samples used. If we make this assumption, then we would all

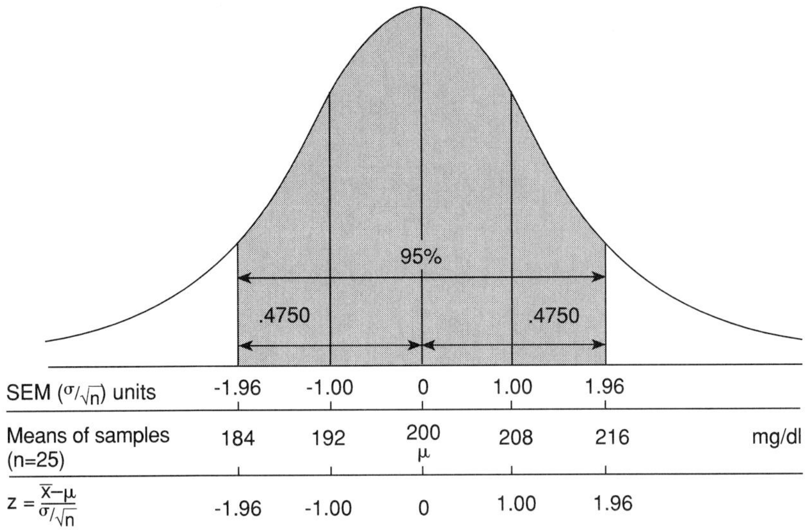


**FIGURE 8.5**
The distribution of sample means (n = 25) from a parent population with a mean (μ) equal to 200 mg/dl and a standard deviation (σ) equal to 40 mg/dl, showing a 95% confidence interval around the population mean (μ).

choose the largest sample (n = 1,600) to make inferences about the population. It makes good sense that we would be able to more accurately estimate population parameters from a sample of 1,600 than from a sample of 16 cases.

## CONFIDENCE INTERVALS

Since we know that the means of repeated samples of the same size are normally distributed, we can look to the normal distribution (see Appendix Table A) to find the probabilities associated with areas under the normal curve as they apply to a distribution of sample means. In Figure 8.5 we have a distribution of sample means in which the population mean (μ) is 200 mg/dl and the SEM ($\sigma/\sqrt{n}$) is 8 mg/dl. Suppose we now want to create the interval that includes 95% of the means of repeated samples (n = 25) taken from the population.

We know that Table A contains the probabilities associated with the areas under the positive half (upper half) of the normal curve. Confidence intervals around population means are divided such that half of the interval is above the population mean and half is below. Therefore, to create a 95% confidence interval we must first divide the interval into halves (95% ÷ 2). We know then that half of the interval, or 47.50% of the area, falls in the positive half, and 47.50% falls in the negative half of the normal curve (Figure 8.5). We can use Appendix Table A to find the z value associated with 47.50% or P = .4750. To do this, we locate P = .4750 in the table and trace our way

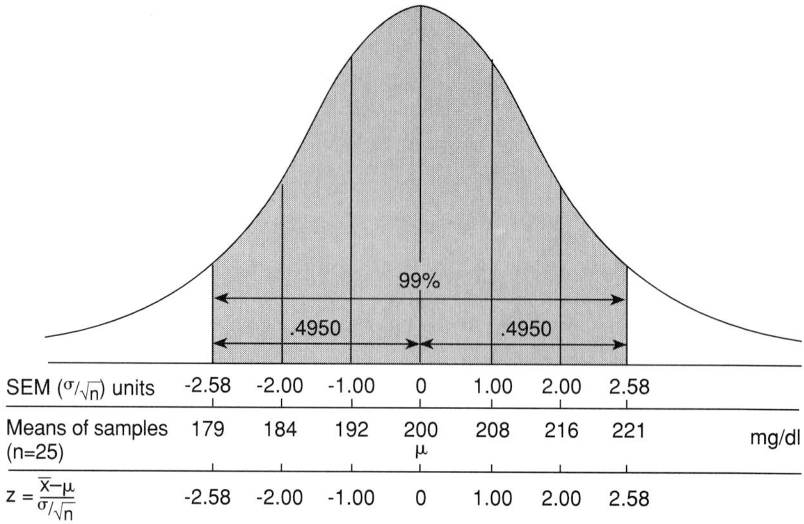


**FIGURE 8.6**
The distribution of sample means (n = 25) from a parent population with a mean (μ) equal to 200 mg/dl total serum cholesterol and a standard deviation (σ) equal to 40 mg/dl, showing a 99% confidence interval around the mean (μ).

down to the X axis to find z = 1.96. In Figure 8.5 the area between z = ±1.96 included 95% of the area under the curve. To find the two sample averages between which 95% of all the means of repeated samples (n = 25) fall, we must calculate the value of the SEM ($\sigma/\sqrt{n}$). Since σ = 40 mg/dl and n = 25, the SEM is equal to $\frac{40 \text{ mg/dl}}{\sqrt{n}}$ = 8 mg/dl. Then we multiply 8 mg/dl by ±1.96. This gives us the number of mg/dl above and below the mean of the population needed to include 95% of all means of samples where n = 25. So, 8 mg/dl multiplied by ±1.96 = ±15.68 mg/dl. For convenience, let us round ±15.68 mg/dl up to 16 mg/dl. Next, add 16 mg/dl to the population mean (μ = 200 mg/dl) and subtract 16 mg/dl from μ to find the upper and lower limits of the 95% confidence interval. These limits are roughly 184 mg/dl and 216 mg/dl (or 200 mg/dl ±16 mg/dl).

Using the same population parameters, let us create a 99% confidence interval around the population mean (μ). First, find the z values associated with a 99% confidence interval. Now divide the interval into halves (99% ÷ 2) to get 49.50% or P = .4950. When we attempt to find the z value associated with P = .4950 (in Appendix Table A) we find we must choose between P = .4949 and P = .4951. If we choose P = .4951, trace down the X axis to find the z value associated with that probability. The z value is 2.58.

In Figure 8.6 we can see that approximately 99% of the sample means fall between z = ±2.58. To find the actual limits, we must calculate the SEM ($\sigma/\sqrt{n}$) as 8 mg/dl and multiply it by ±2.58 to get ±20.64 mg/dl. For convenience, we round up to 21 mg/dl. We then add 21 mg/dl to the population mean ($\mu$ = 200 mg/dl) and subtract 21 mg/dl from $\mu$ to get the interval 179 mg/dl to 221 mg/dl. We can then estimate that 99% of all the means of samples (n = 25) taken from this population will fall between 179 mg/dl and 221 mg/dl, or 200 mg/dl ± 21 mg/dl. Or, if one sample (n = 25) is selected from this distribution, where $\mu$ is 200 mg/dl and $\sigma$ = 40 mg/dl, there is a probability of .99 that its mean ($\overline{X}$) will be between 179 mg/dl and 200 mg/dl. We should also remember that there is a probability of .01 that the sample mean ($\overline{X}$) will be outside (greater than 200 mg/dl or less than 179 mg/dl) this interval.

Although these two examples serve well to demonstrate the creation of confidence intervals, it is unlikely that we would ever be able to select repeated samples from the same population or, in every situation, know the value of the population mean ($\mu$) and standard deviation ($\sigma$). It's more likely we would select only *one* sample and use that sample's statistics as estimates of population parameters.

We estimate population parameters in two different ways. These methods of estimation are the *point estimate* and the *confidence interval estimate*.

When we use the mean of a random sample ($\overline{X}$) to estimate the mean of the parent population ($\mu$) and when the sample standard deviation (s) is used to estimate the population standard deviation ($\sigma$), we are making **point estimates.** A single sample can be randomly selected from a population and its mean ($\overline{X}$) can be used as a point estimate of $\mu$. However, if a second sample of the same size is randomly selected from the same population, the mean of this sample will most likely be different than the mean of the first sample. If we repeatedly took samples of the same size from this population, the means of these samples would form a distribution of sample means, any one of which would be a point estimate of the population mean. Therefore, the problem with point estimation is that it fails to report the precision of the point estimate.

When we know the standard deviation of the population ($\sigma$), two methods are commonly used to indicate the precision of the point estimate of the population mean. The first of these methods is to report the standard error of the distribution of sample means ($\sigma/\sqrt{n}$) with the point estimate ($\overline{X}$) of the population mean. If the mean of a sample of 25 total serum cholesterol values is 202 mg/dl with a SEM of 8 mg/dl, we then could report 202 mg/dl ±8 mg/dl as an estimate of the interval on the sample mean, that, if repeated, would include the mean of the parent population 68% of the time.

The other method is to use *both* the point estimate ($\overline{X}$) and the SEM to create a **confidence interval estimate.** The confidence interval can be used to estimate the probability that other like intervals (created from the means and a standard errors of repeated samples of the same size) would include the mean of the population sampled. Two of the most popular methods used in making inferences regarding the population mean are the 95% and 99% confidence intervals.

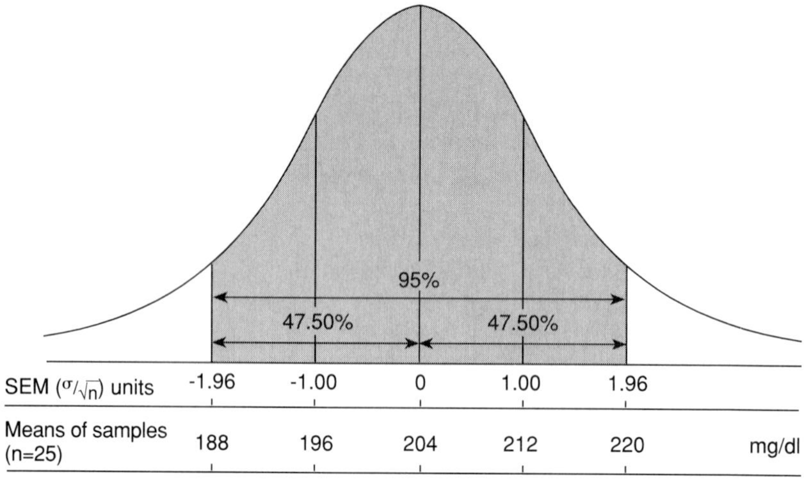


**FIGURE 8.7**
95% confidence interval where: $\overline{X}$ = 204 mg/dl total serum cholesterol, σ = 40 mg/dl, n = 25, and SEM = 8 mg/dl.

## The Creation of the 95% Confidence Interval

The creation of the 95% confidence interval allows us to estimate the width of the interval that (if we repeatedly create such intervals for a large number of samples) would include the mean of the population 95% of the time. Previously we set the mean of the population (μ) as the center of an interval created to include an estimated 95% of the means of repeated samples of the same size taken from the population. Now, because μ is unknown, we set the point estimate and the mean of the sample ($\overline{X}$) at the center of the normal distribution. We then create an interval of sufficient width so that we can conclude there is a .95 probability that our interval is one that includes the mean of the population.

In an effort to estimate the average of a population of total serum cholesterol values, we collect a random sample of 25 values from the population of interest where σ = 40 mg/dl. The sample mean and the standard error estimate are 204 mg/dl and 8 mg/dl, respectively. In Figure 8.7 we set the mean of our sample (204 mg/dl) at the center of the normal distribution. Also see that half of the 95% confidence interval (47.50%) is indicated both above and below the center of the distribution. In Appendix Table A the distance from the center of the distribution we must travel in order to include 47.50% of the area is z = ±1.96. In this case, since we are using the mean of the sample as our point estimate, we know the distance is ±1.96 SEM units. Next, we can find the value of the SEM estimate as $\sigma/\sqrt{n} = 40$ mg/dl $\div \sqrt{25} = 8.0$ mg/dl. Because we know to include ±47.50% of the area under the curve, we must move out from the center of the normal distribution ±1.96 SEM units, then multiply the SEM (8 mg/dl) by ±1.96 to get ±15.86 mg/dl. Again, for convenience, let us round up to 16 mg/dl. In Figure 8.7 the sample mean

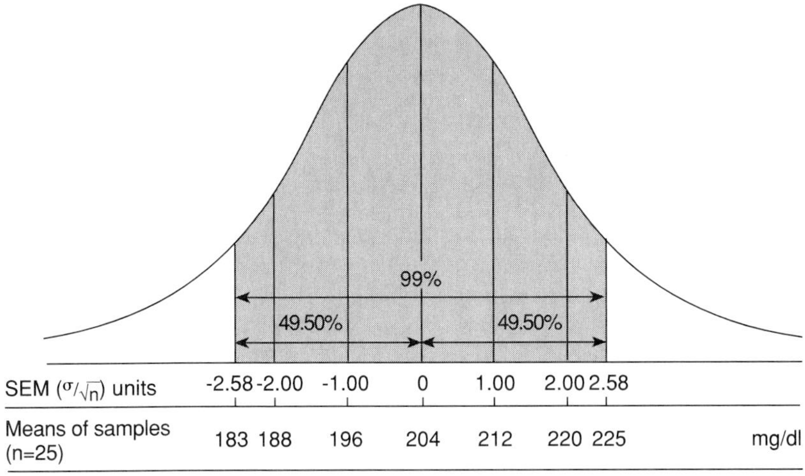


**FIGURE 8.8**
A 99% confidence interval where $\overline{X}$ = 204 mg/dl total serum cholesterol, $\sigma$ = 40 mg/dl, n = 25, and SEM = 8 mg/dl.

(204 mg/dl) ±16 mg/dl creates the interval that extends from 188 mg/dl to 220 mg/dl. We can then conclude that, if we create like intervals from each of many samples of n = 25, 95% of these intervals would include the mean of the population.

## Creation of the 99% Confidence Interval

The creation of the 99% confidence interval allows us to say that, if we apply this process of interval creation to many samples from the population of interest, 99 out of 100 intervals would contain the mean of the population. Using the previous example where the sample mean ($\overline{X}$) = 204 mg/dl and the SEM = 8 mg/dl ($\sigma$ = 40 mg/dl and n = 25), let us create an interval that would give us *more* confidence than P = .95 in our point estimate of the population mean. To achieve an estimate with 99% confidence we know from Appendix Table A that we must move out from the center of the distribution approximately ±2.58 SEM (see Figure 8.8). Since the SEM is equal to 8 mg/dl, 8 mg/dl ×2.58 gives us 20.64 mg/dl. For convenience, round up to 21 mg/dl. Therefore, 204 mg/dl ±21 mg/dl will give us an interval that extends from 183 mg/dl up to 225 mg/dl.

Therefore, we can conclude that this 99% confidence interval we have just created from the sample has a probability of .99 of including the mean of the population.

Compare the ranges of the two intervals presented in Figures 8.7 and 8.8. The range of the 95% interval includes approximately 32 mg/dl (220 mg/dl − 188 mg/dl), while the range of the 99% confidence interval includes approximately 42 mg/dl (225 mg/dl − 183 mg/dl). As the confidence level increases, so do the SEM and thus the range of the confidence interval limits.

The probability expressed in stating confidence intervals with point estimates involves like intervals including the mean of the population, not a statement of probability that the population mean falls in a given confidence interval. Remember, there is but *one* population mean (usually unknown, but having a specific value), while there exists the possibility of *many* like intervals, each having its own unique mean and interval limits. When we create a confidence interval for a single sample, we are estimating the probability that the interval created is one of the intervals that includes the mean of the population.

The 95% and 99% confidence intervals are most commonly reported. Other confidence intervals, such as a 68% or a 90% interval, could be created in the same way.

## SUMMARY

In this chapter we have applied the properties of the Central Limit Theorem to a single sample taken from a parent population. That is, we compared the actual mean ($\overline{X}$) obtained from a sample with the mean ($\mu$) and the standard deviation ($\sigma/\sqrt{n}$ = SEM) of the theoretical distribution of means ($\overline{X}$'s) of repeated samples. This enabled us to determine the probability that the actual mean obtained from the sample ($\overline{X}$) would occur as a result of a random sampling from a parent population with a given mean ($\mu$) and a standard deviation ($\sigma$).

We used the variable

$$z = \frac{(\overline{X} - \mu)}{(\sigma/\sqrt{n})}$$

to compare the actual sample mean ($\overline{X}$) with the expected population mean ($\mu$) by dividing its deviation from $\mu$ ($\overline{X} - \mu$) by its standard deviation. Here, unlike the distribution of individual X values introduced in Chapter 7, we are concerned with the probability distribution of means ($\overline{X}$'s) of samples. Therefore, the standard deviation of the distribution of means is $\sigma/\sqrt{n}$ rather than $\sigma$. Since we know the distribution of means of repeated samples is normal, the same z table (Appendix, Table A) can be used to interpret the values of z and to calculate probability values or percentages of the area under the normal curve.

Whenever we use the normal distribution in the placement of an X value on the $\sigma$ distribution or the placement of an $\overline{X}$ on the $\sigma/\sqrt{n}$ distribution, $\sigma$ must be known. That is, we must know the population standard deviation. In many situations we have no knowledge of the population standard deviation and must estimate $\sigma$ from the sample standard deviation, s. In the next chapter we will discuss another distribution, the t distribution, which is used when $\sigma$ is unknown.

## FOR REVIEW

1. Sampling error is the expected error attributed to chance that we make in selecting random samples to represent populations. Sampling error contributes to inference inaccuracy.
2. Measurement error is caused by mistakes introduced by either the measurer (the person involved in making the measurement) or the method (the measurement tool, device, instrument, test, or questionnaire). Measurement error also contributes to inference inaccuracy.
3. The Central Limit Theorem states that if random samples of size n are repeatedly drawn from any population with a mean $\mu$ and a standard deviation $\sigma$, the distribution of random sample means ($\overline{X}$) will be approximately normally distributed. Also, the mean of the distribution of sample means ($\mu_{\overline{X}}$) will equal $\mu$, and the standard deviation of the distribution of sample means $\sigma_{\overline{X}}$ will equal $\sigma/\sqrt{n}$. Thus, the variable $z = \dfrac{(\overline{X} - \mu)}{\sigma/\sqrt{n}}$ is normally distributed.
4. The parent population is the population that gives rise to a sample or samples.
5. Point estimation occurs when we use the mean of a sample ($\overline{X}$) to estimate the point in the population distribution where the mean of the population ($\mu$) falls. Since we know that other samples of the same size taken from the parent population will have different averages, point estimation alone does not describe the precision of the estimate made. In general, a point estimate is a sample statistic used as an estimate of a population parameter.
6. Often the point estimate is reported with the standard error ($\overline{X} \pm$ SEM). This gives some indication of the precision of the point estimate as the interval indicated by $\overline{X} \pm$ SEM has a probability of about P = .68. Since we know the distribution of sample means (of size n) is normal and that $\sigma_{\overline{X}} = \sigma/\sqrt{n}$, then $\overline{X} \pm \sigma/\sqrt{n}$ includes roughly 68% of the sample means (of size n).
7. The 95% confidence interval is an interval placed on a sample mean taken from a parent population. We can be confident that if such intervals were placed on the means of many repeated samples, then 95% of these intervals would encompass the mean of the population. Thus, the 95% confidence interval on a sample mean is created as

$$\overline{X} \pm 1.96\ (\sigma/\sqrt{n}).$$

8. The 99% confidence interval is an interval placed on a sample mean taken from a parent population. We can be confident that if such intervals were placed on the means of many repeated samples, then 99% of these intervals would encompass the mean of the population. Thus, the 99% confidence interval on a sample mean is created as

$$\overline{X} \pm 2.58\ (\sigma/\sqrt{n}).$$

**Chapter Knowledge and Comprehension Application**

Given a normally distributed population of diastolic blood pressures where $\mu = 80$ mmHg and $\sigma = 3$ mmHg, find:

1. The z score associated with a diastolic pressure of 86 mmHg.
2. The 99% interval of confidence around the sample mean of 84 mmHg, when the sample consists of 25 randomly selected diastolic pressures.
3. The probability that a random sample of 25 diastolic pressures will have an average pressure equal to or less than 81 mmHg.

**Answers**

1. Since we are seeking the z score that is associated with an *individual* X value (a diastolic pressure of 86 mmHg), we use

$$z = \frac{X - \mu}{\sigma},$$

$$= \frac{86 \text{ mmHg} - 80 \text{ mmHg}}{3 \text{ mmHg}},$$

$$= \frac{6 \text{ mmHg}}{3 \text{ mmHg}},$$

$$z = 2.00.$$

2. Use $\overline{X} \pm 2.58\ (\sigma/\sqrt{n}) = 99\%\ CI_{\mu}$

$$84 \text{ mmHg} \pm 2.58\ (3 \text{ mmHg}/\sqrt{25}) =$$

$$84 \text{ mmHg} \pm 2.58\ (3 \text{ mmHg}/5) =$$

$$84 \text{ mmHg} \pm 2.58\ (.6 \text{ mmHg}) =$$

$$84 \text{ mmHg} \pm 1.55 \text{ mmHg} =$$

$$82.45 \text{ mmHg} - 85.\ 55 \text{ mmHg} = 99\%\ CI_{\mu}.$$

3. Since we seek to determine the probability that the mean of a sample of 25 diastolic pressures will be equal to or less than 81 mmHg, we use

$$z = \frac{\overline{X} - \mu}{\sigma/\sqrt{n}},$$

$$= \frac{81 \text{ mmHg} - 80 \text{ mmHg}}{3 \text{ mmHg}/\sqrt{25}},$$

$$= \frac{1 \text{ mmHg}}{3 \text{ mmHg}/5},$$

$$= \frac{1 \text{ mmHg}}{.6 \text{ mmHg}},$$

$$z = 1.67.$$

The probability of z (see Appendix Table A) being equal to, or less than $z = 1.67$ includes the probability of being between $z = 0$ and $z = 1.67$ ($P = .4525$) plus the probability of z being less than $z = 0$ ($P = .5000$).

Therefore, we have a probability of $P = .9525$ that the mean of a randomly selected sample ($n = 25$) will be equal to or less than 81 mmHg.

CHAPTER

# 9

# Estimating the Population Mean Using Student's t Distribution

*"Chance fights ever on the side of the prudent."*

**EURIPIDES**

In the previous chapter we created confidence intervals using the properties of the normal distribution. This required knowledge of the population standard deviation ($\sigma$). Often we do not know and are unable to determine the value of $\sigma$. When $\sigma$ is unknown, we use the standard deviation (s) of a random sample taken from the population as an estimate of $\sigma$. The substitution of the sample standard deviation for the population standard deviation causes a problem. That is, just as we had variability among the means of repeated samples ($\overline{X}_1$, $\overline{X}_2$, $\overline{X}_3$ . . .) taken to estimate the population mean ($\mu$), s also varies from sample to sample. This additional variability among the standard deviations of repeated samples ($s_1$, $s_2$, $s_3$ . . .) is accounted for by the distribution discovered in 1906 by the English statistician William S. Gossett. In 1908 Gossett, who was employed as a statistician and chemist by the Guinness Brewery in Dublin, published his t distribution under the pseudonym "student." The pseudonym was necessary because Guinness, fearing the release of brewery secrets, rarely permitted the printing of employee publications. So, Gossett's distribution is commonly known as **Student's t distribution.**

The main finding of Gossett's work is that when repeated samples are taken from a normally distributed population, the t statistic is very similar to

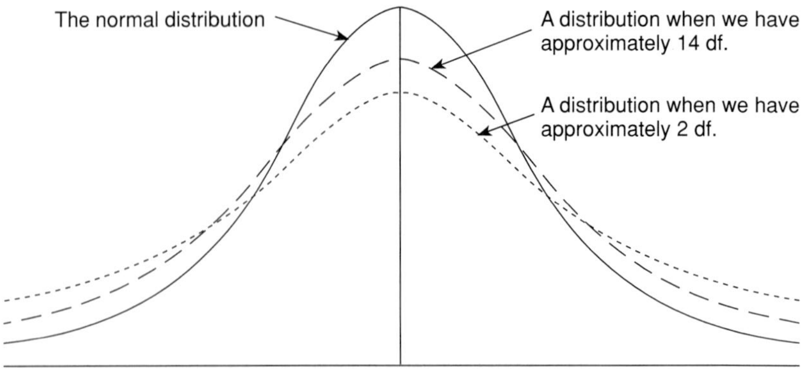


**FIGURE 9.1**
A comparison of the normal z distribution with a t distribution having approximately 14 degrees of freedom (df) and with another t distribution having approximately 2 degrees of freedom.

the z statistic. Both distributions are bell-shaped, symmetrical around $\mu = 0$, extend infinitely in both directions, and have areas under the curve equal to 1.0 or 100%. The main difference between the sampling distributions of the two is that the t distribution is more variable. This increase in variability in the t distribution over the z distribution is understandable since we know that t contains two random quantities ($\overline{X}$ and s), while z has only one ($\overline{X}$). This variability in the sampling distribution of t depends on the size of the samples taken (n). We express this dependency as degrees of freedom $(n - 1)$.

## DEGREES OF FREEDOM

The term **degrees of freedom,** or **df,** is used in conjunction with the t distribution, and with other statistical concepts, to indicate the number of independent observations in a set of observations. In this case (the t distribution), degrees of freedom are related to the computation of the sample standard deviation. When we calculated the sample standard deviation we subtracted the sample mean from each of the individual scores $(X - \overline{X})$. Since the sample mean ($\overline{X}$) is known, specifying *all but one* of these $(X - \overline{X})$ deviations would allow us to know the last deviation. Therefore, $n - 1$ of these deviations $(X - \overline{X})$ are free to vary, and only the *last* one is determined.

Usually the degrees of freedom are equal to the number of cases (n) minus the number of parameters being estimated. When only one parameter is estimated, such as the mean of the population, degrees of freedom (df) equal $n - 1$.

Small sample distributions are more variable than is the normal distribution. The sampling distributions used to help us make statistical decisions when the samples are small tend to be somewhat flatter in the middle and thicker in the tails when compared with the normal distribution. Figure 9.1

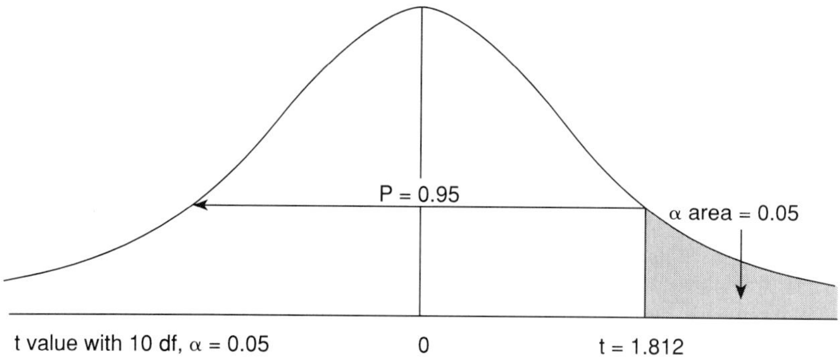


**FIGURE 9.2**
The t value associated with 10 degrees of freedom where area = 0.05.

shows an approximation of the normal distribution and the flatter distributions that occur with small sample sizes. You can see that the distributions become flatter in the middle and thicker at the tails as the degrees of freedom decrease. Also, as the degrees of freedom increase, the t distributions become more like the normal distribution.

These flatter curves indicate an increase in variability as degrees of freedom are lost. Therefore, the properties of the normal distribution do not apply when the distributions of small samples are under consideration. The curves are simply not the same. As we gain degrees of freedom so that the sample's distribution approximates the normal curve, we can apply the properties of the normal distribution. In fact, when we have 30 cases or more, the sample's distribution resembles the normal distribution quite closely. This is because our estimate of $\sigma$, the standard deviation of the sample (s), becomes more precise as the number of sample cases approaches and exceeds 30.

Since the shape of each t distribution depends on the degrees of freedom ($df = n - 1$), there is a different t distribution for each number of df ($n - 1$). To list each possible t value for all possible sample sizes would not be practical. Therefore, Appendix Table B presents the t values that correspond with the upper tail areas ($\alpha$ areas) of 0.10, 0.05, 0.025, 0.01, and 0.005. These upper tail $\alpha$ areas are listed in the left column. The degrees of freedom are presented from left to right along the top and bottom of the table. The t values for each $\alpha$ area are arranged in columns corresponding to the degrees of freedom, from $df = 1$ to $df = 29$. Notice that the corresponding values of t at infinity ($\infty$) are also included.

What t value is associated with a sample size of 11 cases and an alpha area ($\alpha$) of 0.05? First, we calculate the df as $11 - 1 = 10$. Then we can use Table B to find the value of t with 10 df at $\alpha = 0.05$. At the intersection of the column corresponding with an $\alpha$ area of 0.05, we find the t value equal to 1.812.

In Figure 9.2 we can see the t value at $\alpha = 0.05$ with 10 df. Notice the difference between the indication of z and t. In Figure 9.2 we see the t value falls at the point where the $\alpha$ area is to the right of the point. The z value, however, corresponds to the area between 0 and z, as shown in Figure 9.3.

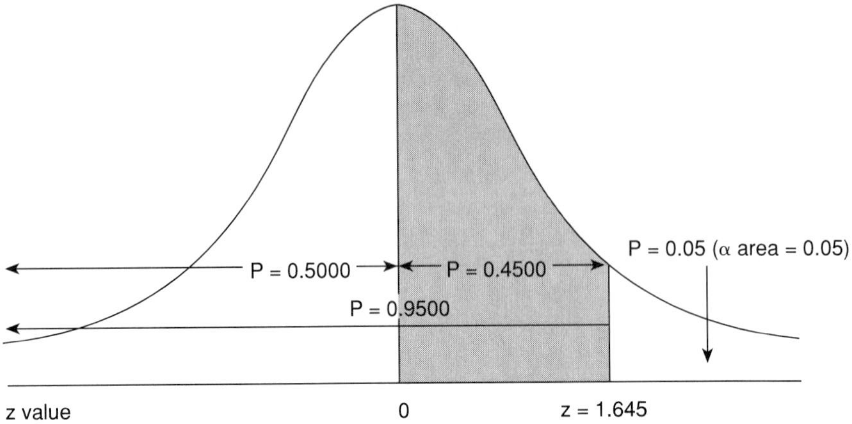


**FIGURE 9.3**
The z value associated with a P = 0.4500 or 45% of the area falling between z = 0 and z = 1.645.

## THE ALPHA (α) AREA AND THE CONFIDENCE COEFFICIENT

We have seen (Appendix Table B) that the location of a particular t value is determined by the degrees of freedom and the alpha area (α). Since the shape (variability) of the t distribution changes for each increase or decrease in sample size (n), especially when n is less than 30, degrees of freedom indicate the appropriate t distribution to use in estimation.

For now, the alpha area (α) and (1 − α) will be described pictorially. Since α is of primary importance in hypothesis testing, we will again discuss α in Chapter 10.

The t values associated with 10 df when the α area is equal to 0.05 are shown in Figure 9.4. Distribution A (the top distribution) is a duplication of Figure 9.2, with one exception: the 1 − α region has been indicated in Figure 9.4a. Since the area under the curve is equal to 1, the area that *does not* include the α area (the unshaded area in Figure 9.4a) is indicated as 1 − α.

The value (1 − α) is the **confidence coefficient.** In Figure 9.4a, all of the α = 0.05 area has been forced into the upper tail of the distribution, leaving the confidence coefficient (1 − α) = 0.95. Remember, we were interested in finding the t value associated with 10 df and an α area of 0.05. With t distributions, the t value falls at the point where the α area is above or to the right of the t value. In this case, if α = 0.05, then (1 − α) = 0.95 and t, with 10 df, is 1.812. Therefore, we can be confident that 95% of the area under the t distribution curve when we have 10 df will fall below a t value of 1.812.

With z, as in Figure 9.3, the shaded area (P = 0.4500) represents the area under the curve between z = 0 and z = 1.645. The α area = 0.05 is the unshaded area above z = 1.645. The (1 − α) area is equal to .95 as well.

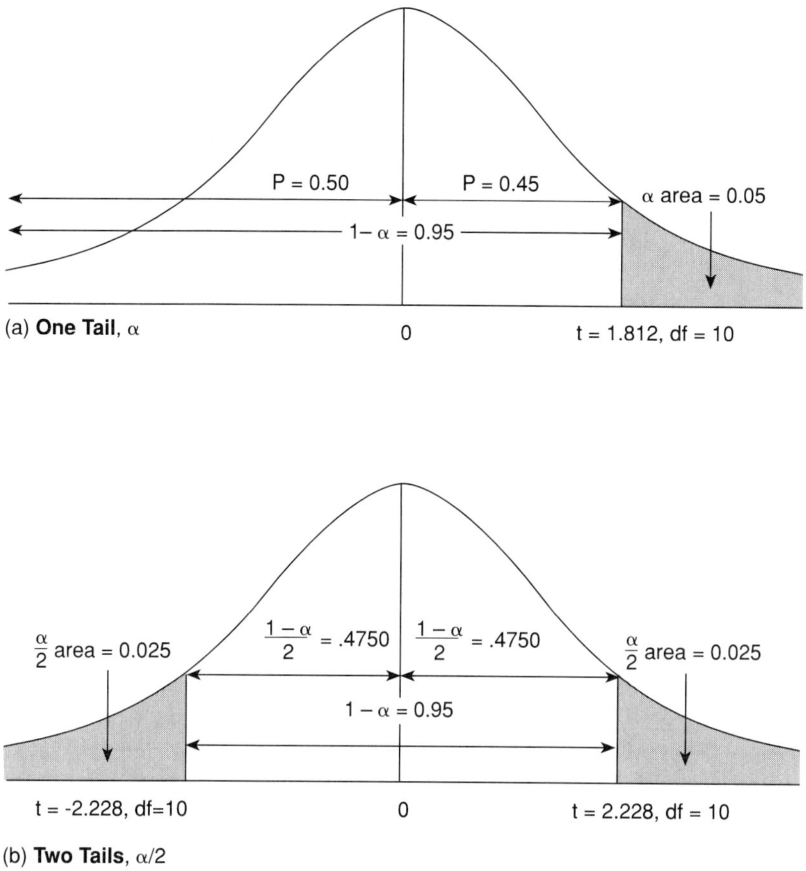


**FIGURE 9.4**
The t values associated with 10 degrees of freedom at (a) $\alpha = 0.05$ when the $\alpha$ region is confined to the upper tail, and (b) when the $\alpha$ region is divided equally between both the upper and lower tails of the distribution.

The level of confidence associated with a confidence interval is also indicated by the value $(1 - \alpha)$. In Figure 9.4b, we can see the interval associated with $1 - \alpha$ confidence includes the middle .95 of the distribution. The area remaining, $\alpha = 0.05$, must be divided equally between each end of the distribution, $\alpha/2$. This area $\alpha/2 = 0.025$ becomes the area to the immediate right of (above) the t value. In this case, with 10 df and using $\alpha/2 = 0.025$ as the area above the t value, t (in Table B) is equal to 2.228. Since the t distributions are symmetrical, we know that the t value associated with the lower tail is $t = -2.228$. We are then confident that 95% of the area under the t distribution curve, when we have 10 df, will be found between $t = \pm 2.228$.

Before we go on to define the t statistic and make inferences about populations based on information obtained from small samples, we will review the properties of the t distributions.

## Properties of the t Distribution

First, the t distribution, like the normal distribution, is symmetrical about $\mu = 0$.

Second, a t distribution's spread or dispersion depends on the number of degrees of freedom in the sample. The standard deviations of the t distributions always exceed 1.0, but decrease to 1.0 as the sample size (hence df) approaches infinity ($\infty$).

Third, the t distributions are more or less bell-shaped (symmetrical around $\mu = 0$). However, as degrees of freedom decrease, the t distribution has higher probabilities of occurrence in the tails and lower probabilities of occurrence in the middle of the distribution than does the normal distribution. As degrees of freedom increase, the t distribution becomes more similar to the shape of the normal distribution. For sample sizes equal to or greater than 30 cases, the t distribution becomes very similar to the normal distribution and the latter can be used for making inferences. At infinity, t and z have equal probabilities.

The change in shape of each t distribution as the df increase is due to the increasing precision of s as a point estimate of $\sigma$ in the calculation of the SEM

$$(s/\sqrt{n}).$$

The sample standard deviation (s) is less accurate as an estimate of $\sigma$ as degrees of freedom become fewer, thus increasing sampling error. When the sampling error is greatest, in the small sample case, the t sampling distribution becomes more dispersed than the normal sampling distribution.

Fourth, since each t distribution has a slightly different shape (therefore, different probabilities for each distinct degree of freedom), a completely different set of t values is needed for each degree of freedom value.

## The t Statistic

The **t statistic** may be defined as

$$t = \frac{\overline{X} - \mu}{s/\sqrt{n}} \quad \text{where,}$$

$\overline{X}$ is the mean of a randomly selected sample of n size when n is equal to or less than 30;

s is the standard deviation of the randomly selected sample;

$(s/\sqrt{n})$ is the estimate of the standard error of the mean, and
$\mu$ is equal to the mean of the population from which the sample was randomly selected.

When the mean of the population ($\mu$) is unknown we can use the t statistic to estimate the mean of the population of interest.

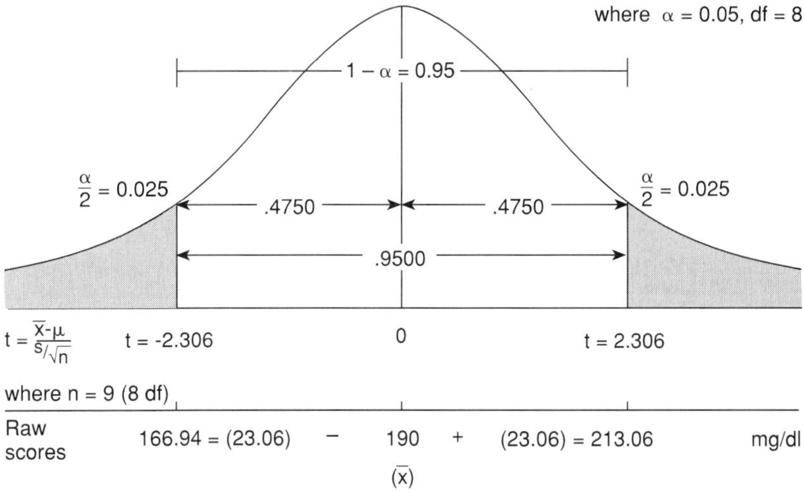


**FIGURE 9.5**
The t values for a 95% confidence interval when n = 9 (9 − 1 = 8 df), s = 10 mg/dl.

## THE 95% CONFIDENCE INTERVAL USING THE t STATISTIC

If a random sample of nine healthy adult women was found to have an average total serum cholesterol level of 190 mg/dl and a standard deviation of 30 mg/dl, what would be the 95% confidence interval for the mean total serum cholesterol level of the population of healthy adult women? In this case we can use the sample mean ($\overline{X}$ = 190 mg/dl) as the point estimate of the mean of the population. Next, the t values that correspond to a 95% confidence interval must be determined. That is, we must find the t values for the interval that has a probability of .95 of being inclusive of the true mean of the population.

We know the shape (variability) of the t distribution depends upon the number of degrees of freedom. In this case, when estimating μ, we have a sample with n − 1 degrees of freedom, or 9 − 1 = 8 df. We divide the interval equally on either side of the point estimate ($\overline{X}$ = 190 mg/dl), which we have placed at the center of the t distribution with 8 df (Figure 9.5).

We know that, unlike z, t values are associated with the point where the area just above or to the right of the t value (the shaded area) is the α area. In this case the 0.05 α area must be divided equally into each tail (α/2 = 0.25). We know that this area to the right of t is *half* of the alpha area. We know it is not α, yet it is the area just above or to the right of the t value. So, instead of using the α = 0.05 row in Table B, we think of this half of the confidence interval as being a single tail where α = 0.025. Therefore, the t value at 10 df with 0.025 area to the right is t = 2.306. Likewise, because of the symmetrical nature of the t distributions, t = −2.306 has 0.025 of the area below it. This small adjustment in our thinking allows us to use Table B without having to add several rows of values for t when the α area is split into two tails, as with the confidence interval.

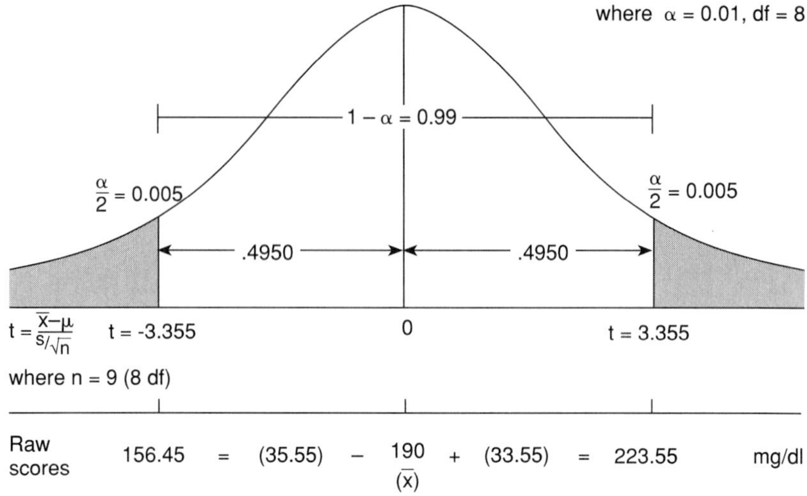


**FIGURE 9.6**
The t values for a 99% confidence interval when n = 9 (9 − 1 = 8 degrees of freedom), $\overline{X} = 190$ mg/dl, s = 30 mg/dl, and $s/\sqrt{n} = 10$ mg/dl.

Now that we know the t values, we can multiply them both by the estimate of the standard error of the sampling distribution, where n = 9 and s = 30 mg/dl. These two values can then be added to the point estimate of μ, $\overline{X} = 190$ mg/dl. Therefore, we have

$$\overline{X} \pm t_{(\alpha/2\,=\,.025,\ 8\ \text{df})}\,(s/\sqrt{n}) = 95\%\ CI_{\mu},$$
$$190\ \text{mg/dl} \pm 2.306\ (30\ \text{mg/dl}\sqrt{9}) =$$
$$190\ \text{mg/dl} \pm 2.306\ (10\ \text{mg/dl}) =$$
$$190\ \text{mg/dl} \pm 23.06 =$$
$$[166.94\ \text{mg/dl to } 213.06\ \text{mg/dl}] = 95\%\ CI_{\mu}.$$

We can conclude that for every 100 confidence intervals created in this way, 95 of them will include the mean of the population, μ.

## THE 99% CONFIDENCE INTERVAL USING THE t STATISTIC

If a random sample of nine healthy adult women was found to have an average total serum cholesterol level of 190 mg/dl and a standard deviation of 30 mg/dl, what would be the 99% confidence interval for the mean serum cholesterol level of the population of healthy adult women? Since $1 - \alpha = 0.99$, $\alpha = 0.01$. Since we must equally divide α between the lower and upper tails of the distribution, $\alpha/2 = 0.005$ (Figure 9.6). Therefore, the t value for 8 df (9 − 1) with P = 0.005 just above t is 3.355. Thus, we have

$$\overline{X} \pm t_{(\alpha/2\,=\,.005,\ 8\ df)}\,(s/\sqrt{n}) = 99\%\ CI_{\mu},$$
$$190\ mg/dl \pm 3.355\ (30\ mg/dl\sqrt{9}) =$$
$$190\ mg/dl \pm 3.355\ (10\ mg/dl) =$$
$$190\ mg/dl \pm 33.55 =$$
$$[156.45\ mg/dl\ to\ 223.55\ mg/dl] = 99\%\ CI_{\mu}.$$

We can conclude that if we create many, many intervals in this way, 99% of them will include the mean of the population, $\mu$.

## SUMMARY

The t distribution is a symmetrical distribution about a mean of 0. For each increase or decrease in sample size, hence degrees of freedom, the standard deviation of the t distribution changes. The standard deviation of t always exceeds 1.0 but approaches 1.0 as the sample size increases. Therefore, for each sample size there is a different t distribution.

We use the t distribution when we have small sample sizes such that n is less than 30 cases. Since we use s, the sample standard deviation, to estimate $\sigma$, we use t when inferences are required concerning populations where $\sigma$ is unknown.

## FOR REVIEW

1. A t value is the difference between the mean of a population ($\mu$) and the mean of a random sample ($\overline{X}$) taken from the population, divided by the estimate of the standard error of the sampling distribution [$(s/\sqrt{n})$] where s is an estimate of unknown $\sigma$. Therefore, the t value for the sample mean $\overline{X}$ is

   $t = \dfrac{\overline{X} - \mu}{s/\sqrt{n}}$. Note, if $\overline{X} - \mu$ is equal to $(s/\sqrt{n})$, then $t = 1.00$; if $\overline{X} = \mu$, then $t = 0$.

2. The degrees of freedom (df) are an indication of the reliability of s, in that the larger the size of the sample represented by s, the more reliable s will be as an estimate of $\sigma$.
3. The confidence coefficient is equal to perfect probability ($P = 1.00$) minus the alpha ($\alpha$) area $(1 - \alpha)$, where the alpha area ($\alpha$) corresponds to the probability of all samples in the sampling distribution that would not include the estimated parameter of interest, in this case, $\mu$. If $\alpha = 0.05$ then $1 - \alpha = .95$ indicating, for example, that we have a probable level of confidence of .95 in our estimate of the parameter. The smaller the $\alpha$, the larger the probability of confidence.

4. The customary form of a confidence interval on $\overline{X}$ to estimate $\mu$ is

$$CI_\mu = \overline{X} \pm t_{(\alpha/2,\ df)} (s/\sqrt{n}).$$

At 10 df, the 95% confidence interval becomes

$$95\%\ CI_\mu = \overline{X} \pm t_{(0.05/2,\ 10)} (s/\sqrt{n}),$$
$$95\%\ CI_\mu = \overline{X} \pm 2.228(s/\sqrt{n}).$$

At 10 df, the 99% confidence interval becomes

$$99\%\ CI_\mu = \overline{X} \pm t_{(0.01/2,\ 10)} (s/\sqrt{n}),$$
$$99\%\ CI_\mu = \overline{X} \pm 3.169\ (s/\sqrt{n}).$$

**Chapter Knowledge and Comprehension Application**

Given: A random sample of 25 men is selected from the population of active adult males participating in a marathon race. The sample's average hemoglobin value is 16 g/100 ml of blood with a standard deviation of 2 g/100 ml of blood.

1. Create a 95% confidence interval for the mean of the population and supply its interpretation.
2. Create a 99% confidence interval for the mean of the population and supply its interpretation.

**Answers**

1. $CI_\mu = \overline{X} \pm t_{(\alpha/2,\ df)} (s/\sqrt{n})$
   (We use t here because $\sigma$ is unknown and n is less than 30.)

$$\begin{aligned} 95\%\ CI_\mu &= 16\text{g}/100\text{ ml} \pm t_{(0.05/2,\ 24)} \frac{(2\text{g}/100\text{ ml})}{\sqrt{25}}, \\ &= 16\text{ g}/100\text{ ml} \pm t_{(0.025,\ 24)} \frac{(2\text{ g}/100\text{ ml})}{5}, \\ &= 16\text{ g/ml} \pm 2.064\ (0.4\text{ g}/100\text{ ml}), \\ &= 16\text{ g/ml} \pm 0.826\text{ g}/100\text{ ml}, \\ 95\%\ CI_\mu &= [15.174\text{ g}/100\text{ ml to }16.826\text{ g}/100\text{ ml}]. \end{aligned}$$

*Interpretation:*

For every 100 intervals created in this way, using the means and standard deviations of random samples where n = 25, we can expect 95 of such intervals to include the true mean of the parent population, $\mu$.

2. $CI_\mu = \overline{X} \pm t_{(\alpha/2,\ df)} (s/\sqrt{n})$.
   (We use t here because $\sigma$ is unknown and n is less than 30.)

$$99\%\ CI_{\mu} = 16\text{g}/100\text{ ml} \pm t_{(0.01/2,\ 24)} \frac{(2\text{g}/100\text{ ml})}{\sqrt{25}},$$

$$= 16\text{ g}/100\text{ ml} \pm t_{(0.005,\ 24)} \frac{(2\text{ g}/100\text{ ml})}{5},$$

$$= 16\text{ g}/100\text{ ml} \pm 2.797\ (0.4\text{ g}/100\text{ ml}),$$

$$= 16\text{ g}/100\text{ ml} \pm 1.119\text{ g}/100\text{ ml},$$

$$99\%\ CI_{\mu} = [14.881\text{ g}/100\text{ ml to } 17.119\text{ g}/100\text{ ml}].$$

*Interpretation:*

For every 100 intervals created in this way—using the means and standard deviations of random samples where n = 25—we can expect 99 of such intervals to include the true mean of the parent population, μ.

CHAPTER

# 10

# Single Samples and Tests of Hypotheses About the Population Mean

*"By a small sample we may judge of the whole piece."*

**MIGUEL DE CERVANTES**
**1547–1616**

In Chapter 1 we learned that researchers have attempted to organize their inductive research efforts by using the scientific method. The scientific method involves observing data, formulating a hypothesis, designing an experiment, and interpreting the results in a way that either supports or rejects the formulated hypothesis. In this chapter we will formulate hypotheses, test them, and make conclusions as indicated by the tests' results. We will focus on statistical hypothesis tests concerning the mean of a single population. These tests will adhere to the following general steps:

1. A hypothesis concerning the value of the mean of a single population will be formed.
2. The difference between the observed mean of a single sample and the hypothesized value of the population mean will be divided by the standard error to determine the probable "size" of the difference.
3. The hypothesis concerning the mean of a single population being equal to a certain value will be rejected or not rejected depending on the probability associated with the observed mean of the sample.

Now we will consider some specific components for each of the three general steps for testing hypotheses concerning the means of single populations.

## FORMING THE HYPOTHESES

Two hypotheses are stated during statistical hypothesis testing: the null hypothesis and the alternative hypothesis. The **null hypothesis ($H_0$:)** is always formed as a statement of *statistical* equality. That is, the null hypothesis does not state that the mean of a population equals some value *exactly,* but that the sample mean (observed) and the hypothesized population mean are not different enough to reject the conclusion that they represent the same population. Therefore, the null hypothesis is the hypothesis of no *significant* statistical difference; it is a convenience that allows for the calculation of statistical probability. We set the mean of the population equal to a hypothesized value as if we were confident of our knowledge of the true sampling distribution. If we reject the null hypothesis, we claim a significant statistical difference that is not due to chance alone. If we fail to reject the null hypothesis, the difference is *not significant* statistically and therefore is attributable to chance sampling error.

The researcher usually has no expectation that the null hypothesis is true. In fact, the researcher sets an opposing alternative or research hypothesis so that the null hypothesis might be rejected. A rejected null hypothesis does not suggest failure on the part of the researcher. The rejection refers only to the researcher's decision that the chance of the null hypothesis being true is remote. A rejected null hypothesis indicates a significant outcome. If we fail to reject the null hypothesis, the indication is that the amount of difference is due to chance and is not statistically significant. An example of a null hypothesis that might be used in tests concerning the mean of a single population is:

$H_0$: The population mean ($\mu$) is equal to some hypothesized value ($\mu_H$),

$H_0$: $\mu = \mu_H$.

The **alternative hypothesis ($H_a$:)** is the research or test hypothesis. Therefore, the alternative hypothesis is formed so that it corresponds with the ultimate goal of the experiment. The alternative hypothesis is a statistical restatement of the research question, an expression of the theory that we hope to find evidence to support. The alternative hypothesis opposes the null so that whenever we reject the null hypothesis we support the alternative. Likewise, whenever we fail to reject the null, we are not in support of the alternative.

The alternative hypothesis can oppose the null in three ways. First, if the research theory is that a difference exists between the hypothetical population mean and the observed mean of a sample, then a hypothesis of *difference* is formed. An example of an alternative hypothesis of difference that opposes the null hypothesis—$H_0$: the population mean ($\mu$) is equal to some hypothesized value ($\mu_H$)—may be written as

$H_a$: The population mean ($\mu$) is different than the hypothesized value ($\mu_H$), or $H_a$: $\mu \neq \mu_H$.

This alternative hypothesis of difference is a nondirectional or two-sided hypothesis because it does not specify that the population mean ($\mu$) is greater than (>) the value ($\mu_H$) or that the population mean is less than (<) the value. So, *either* a greater-than (>) *or* less-than (<) difference that is not due to chance alone could lead to the rejection of the null hypothesis in favor of the alternative hypothesis of difference.

Second, research theory sometimes indicates that the population mean is greater than the hypothesized value ($\mu_H$). In this case, we form a one-sided (greater-than) alternative hypothesis, an example of which may be written as

$H_a$: The population mean ($\mu$) is greater than the hypothesized value ($\mu_H$), or $H_a$: $\mu > \mu_H$.

This greater-than alternative hypothesis is directional, or one-sided.

The third $H_0$-opposing alternative hypothesis is the one-sided, less-than hypothesis. It may be expressed as

$H_a$: The population mean ($\mu$) is less than the hypothesized value ($\mu_H$), or $H_a$: $\mu < \mu_H$.

When *one* of these last two directional hypotheses are indicated, we must remember that the opposing null hypothesis would in fact necessarily include the other direction. For example, if the alternative hypothesis indicated is $H_a$: $\mu > \mu_H$, then the opposing null must indicate $H_0$: $\mu \leq \mu_H$. That is, if the alternative is that $\mu$ is greater than $\mu_H$, then the null hypothesis indicates that $\mu$ is equal to or less than $\mu_H$ by default. If the alternative is that $\mu$ is less than $\mu_H$, then the null indicates that $\mu$ is equal to or greater than $\mu_H$.

## SELECTING THE TEST STATISTIC

The **test statistic** is used to test the difference between the observed mean of a single sample and the hypothesized value of the population mean. One of two, one-sample test statistics will be calculated to help us decide whether to reject the null hypothesis. These are the one-sample z test statistic and the one-sample t test statistic. These are expressed as

$$z = \frac{\overline{X} - \mu_H}{\sigma/\sqrt{n}} \quad \text{and} \quad t = \frac{\overline{X} - \mu_H}{s/\sqrt{n}}.$$

We can determine which of these test statistics to use simply by recognizing three possible situations. First, if the population standard deviation ($\sigma$) is unknown and the sample size is small ($n \leq 30$) then the t test statistic is used. In this case, the sample standard deviation (s) is calculated and used as an estimate of $\sigma$. Second, if the population standard deviation is known, the z test statistic is used. Third, when $\sigma$ is unknown and the sample size is large, it usually makes little difference whether the z or t test statistic is used. Remember, when the sample size is very large, the z and t values associated with certain alpha ($\alpha$) values become very similar.

## DETERMINING THE NULL HYPOTHESIS REJECTION REGION

The **null hypothesis rejection region** is selected so that its probability of containing the numerical value of the test statistic, if the null hypothesis is true, equals alpha ($\alpha$). In other words, the rejection region is the region defined by $\alpha$ that represents the probability of rejecting the null hypothesis when the null hypothesis is, in fact, true. Since this rejection region, represented by $\alpha$ probability, is the chance we have of rejecting a true null hypothesis (making an $\alpha$ error), we usually select small $\alpha$ error probabilities such as $\alpha = 0.05$ or $\alpha = 0.01$. This $\alpha$ error, the chance of rejecting the null hypothesis when the null is true, is also called a **Type I error.**

By setting the $\alpha$ region at 0.05, for example, we not only define the null hypothesis rejection region but we simultaneously define the region where we fail to reject the null hypothesis. This null hypothesis nonrejection region is defined as $1 - \alpha$ (if $\alpha = 0.05$ then $1 - \alpha = 0.95$). The area associated with our failure to reject the null hypothesis ($1 - \alpha$) is increased as the null hypothesis rejection region ($\alpha$) is reduced. Just as the $\alpha$ value represents our chance of erroneously *rejecting* a *true* null hypothesis, there is also an error associated with the acceptance of a false null hypothesis. The probability of *accepting* a *false* null hypothesis is called the **beta ($\beta$)** or **Type II error.** Even though we usually do not know the value of $\beta$, there is always some chance of making a $\beta$ error. Therefore, if the calculated test statistic value falls in the rejection region, we can reject the null hypothesis and know that the probability of rejecting the null when it is true is equal to $\alpha$. If the test statistic value does not fall in the rejection region, we do not *accept* the null hypothesis unless we know $\beta$. That is, we *fail to reject* the null hypothesis in order to withhold judgment as to the truth or falsity of the null hypothesis. We could then seek a larger sample and retest or estimate $\mu$ by creating a confidence interval.

## MAKING A CONCLUSION

If the calculated test statistic value falls in the rejection region, we *reject* the null hypothesis and conclude that the alternative hypothesis is true, keeping in mind that the chance of erroneously rejecting a true null hypothesis is equal to $\alpha$. If the calculated test statistic value falls outside the rejection region, then we *fail to reject* the null hypothesis. We do not accept the null unless we know the value of $\beta$. Our failure to reject the null is an indication that we are unsure about which hypothesis is true.

Now let us look at some examples of one-sample hypothesis tests based on the following information: Lithium carbonate is a drug prescribed for the treatment of psychiatric patients who are diagnosed with Bipolar Affective Disorder (BPAD). A therapeutic blood level of lithium carbonate equals 0.9 to 1.4 mEq/L. Blood levels of 1.5 mEq/L or greater are considered toxic, and levels of 2.0 mEq/L or greater are potentially lethal.

## The One-sample Test of a Hypothesis About a Population Mean When $\alpha$ is Known

Suppose the average lithium maintenance level among the population of all BPAD psychiatric patients in a large health-care facility is reported to be 1.3 mEq/L with a standard deviation of 0.3 mEq/L. Does a randomly selected sample of 25 patients, having a mean lithium carbonate maintenance level of 1.4 mEq/L, provide evidence that the average maintenance level for the population of BPAD psychiatric patients at the facility is significantly *greater* than the reported 1.3 mEq/L? Use $\alpha = 0.01$.

### Example solution:

#### *Step 1 Forming the hypotheses:*

$$H_0: \mu \leq 1.3 \text{ mEq/L},$$

$$H_a: \mu > 1.3 \text{ mEq/L}.$$

The one-sided alternative hypothesis is indicated by the research question. Namely, is there evidence that the average maintenance level for the population of BPAD psychiatric patients is greater than 1.3 mEq/L? The null, in this case, is that the average maintenance level is 1.3 mEq/L or less.

#### *Step 2 Selecting the test statistic:*

$$z = \frac{\overline{X} - \mu}{\sigma/\sqrt{n}},$$

$$= \frac{1.4 \text{ mEq/L} - 1.3 \text{ mEq/L}}{0.3 \text{ mEq/L} \div \sqrt{25}} = 1.67.$$

The one-sample z test is indicated here because we know that $\mu = 1.3$ mEq/L and $\sigma = 0.3$ mEq/L. Also, only one sample, having $\overline{X} = 1.4$ mEq/L, is randomly selected from the population. The calculated z value is 1.67.

#### *Step 3 Determining the null hypothesis rejection region:*

Since we have a directional, one-sided alternative hypothesis with $\alpha = 0.01$, our critical z value ($z = 2.33$) is indicated in Figure 10.1.

From Appendix Table A we find that we have a probability of approximately .4900 (or a 49% chance) of having a z score between $z = 0$ and

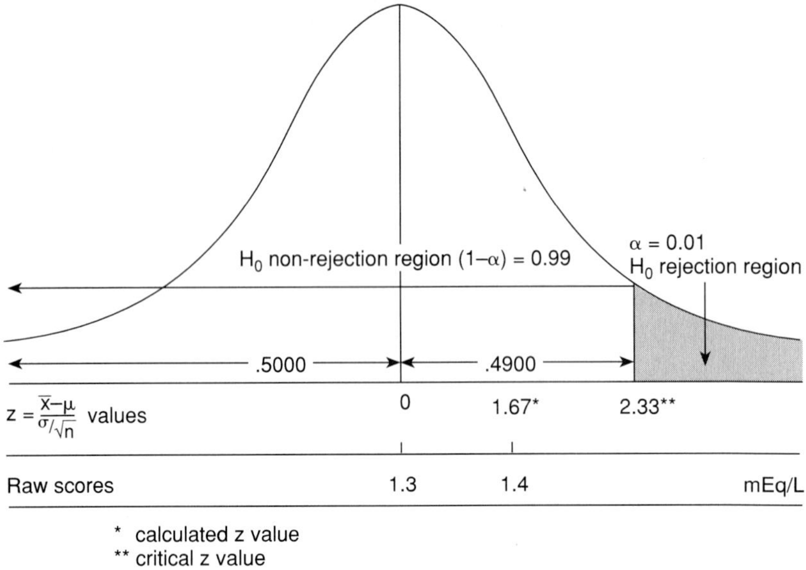


**FIGURE 10.1**
The critical z value (2.33) associated with an alpha region of 0.01 and a greater-than, directional test of the null hypothesis.

z = 2.33. Thus, z = 2.33 is the critical z value. In this case, the critical z value is the value above which approximately 0.01, or 1% of the cases fall and below which .99, or 99% (50% + 49%) of the cases fall. If we wish to allow only a 1% chance of error ($\alpha = 0.01$) we will only reject the null hypothesis if the calculated z value equals or is greater than the $\alpha = 0.01$ critical z value of 2.33.

### *Step 4 Making a conclusion:*

The calculated z value (z = 1.67) that corresponds with the sample mean of 1.4 mEq/L is shown in Figure 10.1. Notice that this calculated z value of 1.67 *does not* fall in the null hypothesis rejection region (area defined by $z \geq 2.33$). Therefore, we fail to reject the null hypothesis. We can conclude that the sample average ($\overline{X} = 1.4$ mEq/L) was not large enough for us to be 99% confident that the average lithium carbonate maintenance level for the population was greater than 1.3 mEq/L.

We could then retest, after selecting a larger random sample, or we could create a 99% confidence interval around the sample mean such that

$$
\begin{aligned}
99\%\ \mathrm{CI}_{\mu} &= \overline{X} \pm z_{(.4950)}\,(\sigma\sqrt{n}),\\
&= 1.4\ \mathrm{mEq/L} \pm 2.58\left(\frac{0.3\ \mathrm{mEq/L}}{\sqrt{25}}\right),\\
&= 1.4\ \mathrm{mEq/L} \pm .15\ \mathrm{mEq/L},\\
99\%\mathrm{CI}_{\mu} &= [1.25\ \mathrm{mEq/L\ to\ } 1.55\ \mathrm{mEq/L}].
\end{aligned}
$$

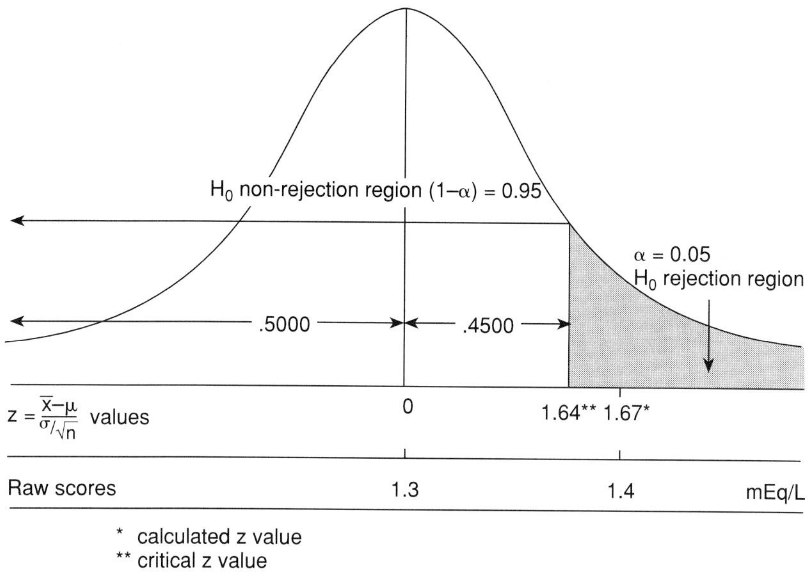


**FIGURE 10.2**
The critical z value (1.64) associated with an alpha region of 0.05 and a greater-than, directional test of the null hypothesis.

Thus, we are confident that the true population mean ($\mu$) will be included in 99% of such intervals.

Notice that the reported population mean 1.3 mEq/L falls in the lower end of our 99% confidence interval. What if the original alternative hypothesis ($H_a$: $\mu > 1.3$ mEq/L) was tested at $\alpha = 0.05$ instead of $\alpha = 0.01$? This, in effect, allows for an increased chance of rejecting the true null hypothesis and a decreased level of confidence (from .99 to .95). The result would have been the same for Steps 1 and 2 of the test procedure. For Step 3, the null hypothesis rejection region would have been indicated as the area above a critical z value of 1.64. In Figure 10.2 we can see that the calculated z value (1.67) falls within the null hypothesis rejection region indicated as the $\alpha$ area above the critical z of 1.64.

In this case we would reject the null hypothesis and conclude, accepting a 5% chance of error, that the average lithium carbonate maintenance level for the population is greater than 1.3 mEq/L, as reported. As you can see, if we are willing to accept a greater chance of error (by increasing the $\alpha$ area) the likelihood of rejecting the null hypothesis is increased. When the calculated value of the test statistic falls in the null hypothesis rejection region, it does so only rarely, as indicated by the small area of $\alpha$ probability (when $\alpha = 0.05$, 0.01 or less). Such events are deemed unusual or statistically significant. Therefore, when a test statistic falls in the null hypothesis rejection region, the event is referred to as significant at the particular level of alpha ($\alpha$).

## The One-sample Test of a Hypothesis About a Population Mean When σ is Unknown

Suppose, in this case, that the average lithium maintenance level among the population of all Bipolar Affective Disorder (BPAD) psychiatric patients in a large health-care facility is reported to be 1.3 mEq/L. Imagine a randomly selected sample of nine patients, having an average lithium carbonate maintenance level of 1.4 mEq/L with a standard deviation of 0.3 mEq/L. Does this sample provide evidence that the average maintenance level for the population of all BPAD psychiatric patients at the facility is not equal to 1.3 mEq/L? Use $\alpha = 0.05$.

### Example solution:

***Step 1 Forming the hypotheses:***

$$H_0: \mu = 1.3 \text{ mEq/L},$$

$$H_a: \mu \neq 1.3 \text{ mEq/L}.$$

The two-sided alternative hypothesis is indicated by the research question. Namely, is there evidence that the average maintenance level for the population of all BPAD psychiatric patients is *not equal to* 1.3 mEq/L? The null, in this case, is that the average maintenance level is 1.3 mEq/L.

***Step 2 Selecting the test statistic:***

$$t = \frac{\overline{X} - \mu}{s/\sqrt{n}},$$

$$t = \frac{1.4 \text{ mEq/L} - 1.3 \text{ mEq/L}}{0.3 \text{ mEq/L} \div \sqrt{9}},$$

$$t = 1.00$$

The one-sample t test is indicated because we do not know the value of the population's standard deviation (σ).

Therefore, the standard deviation of the sample ($s = 0.3$ mEq/L) is used as an estimate of σ. Also, only one sample is randomly selected from the population. The calculated t value is 1.00.

***Step 3 Determining the null hypothesis rejection region:***

Since the alternative hypothesis is a nondirectional, two-sided hypothesis, and since the α region was set at 0.05 ($\alpha/2 = 0.025$) with 8 df ($n - 1$), our critical t values are ±2.31 (Appendix Table B). These critical t values ($\alpha/2 = 0.025$, 8 df) are indicated in Figure 10.3, along with the null hypotheses rejection regions. Our critical t values ($\alpha/2 = 0.025$, 8 df) define the limits of the null hypothesis rejection regions as ±2.31. The calculated t value of 1.00 does not fall in the null hypothesis rejection region.

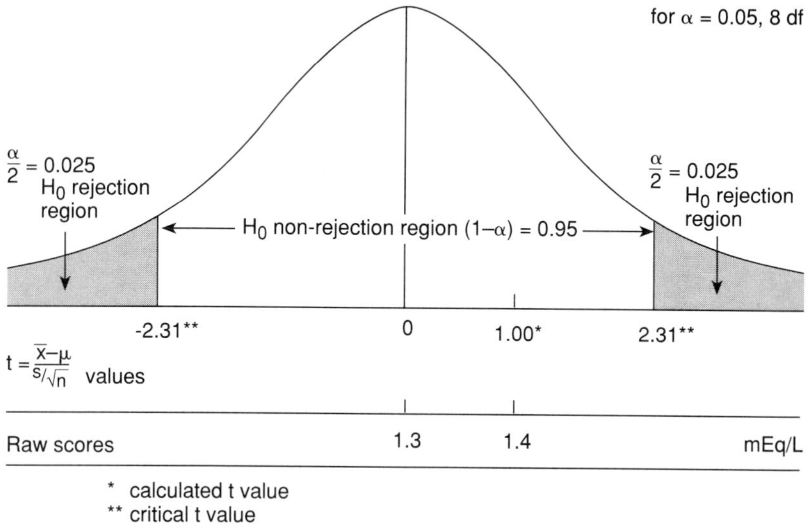


**FIGURE 10.3**
The critical t values (±2.31) associated with $\alpha = 0.05$, 8 degrees of freedom, and a nondirectional, two-sided test of the null hypothesis.

### *Step 4 Making a conclusion:*

Since our calculated t value ($t = 1.00$) is not equal to or greater than the critical t values (±2.31), we fail to reject the null hypothesis. We conclude that the sample average ($\overline{X} = 1.4$ mEq/L) is not different enough from the reported population mean ($\mu = 1.3$ mEq/L) to be 95% confident that the average lithium carbonate maintenance level for the population is not equal to 1.3 mEq/L.

We could retest the hypothesis after selecting a larger sample, or we could create a 95% confidence interval around the sample mean such that

$$\begin{aligned} 95\%CI_{\mu} &= \overline{X} \pm t_{(\alpha/2\,=\,0.025,\ 8\ df)}\,(\sigma/\sqrt{n}), \\ &= 1.4\ \text{mEq/L} \pm 2.31\ (0.3\ \text{mEq/L} \div \sqrt{9}), \\ &= 1.4\ \text{mEq/L} \pm .231, \\ 95\%CI_{\mu} &= [1.17\ \text{mEq/L to } 1.63\ \text{mEq/L}]. \end{aligned}$$

Thus, we are confident that 95% of the intervals constructed in this way will include the true mean of the population.

Remember, to use the t test statistic for one sample situation, some simple requirements must be met:

1. Only one sample has been randomly selected from the population of interest.
2. The characteristics being measured (the raw scores) must not depart significantly from normality within the population.
3. The characteristics being measured (the raw scores) must be interval data or ratio data.

## SUMMARY

In this chapter we have considered single-sample tests of hypotheses about the means of populations. A summary of these tests follows:

1. The one-sample situation where the sample is randomly selected from the population and $\sigma$ *is known:*
   a. The null hypothesis is stated such that the population mean ($\mu$) is equal to some hypothesized or observed value ($\mu_H$). Therefore, $H_0$: $\mu = \mu_H$.

   b. The alternative may be, depending on the research hypothesis, one of three possibilities:
   $H_a$: $\mu \neq \mu_H$ (nondirectional, two-sided alternative),
   $H_a$: $\mu > \mu_H$ (directional, one-sided, greater-than alternative), or
   $H_a$: $\mu < \mu_H$ (directional, one-sided, less-than alternative).
   c. The test statistic:

   $$z = \frac{\overline{X} - \mu_H}{\sigma/\sqrt{n}} \qquad \text{(the one-sample z test statistic)}$$

   d. $H_0$: Rejection regions:
   For $\alpha = 0.05$, when $H_a$ is directional, the critical $z =$ either $+1.645$ or $-1.645$.
   For $\alpha = 0.05$, when $H_a$ is nondirectional, the critical $z = \pm 1.96$ $[z(1 - \alpha)/2]$.
   For $\alpha = 0.01$, when $H_a$ is directional, the critical $z =$ either $+2.33$ or $-2.33$.
   For $\alpha = 0.01$, when $H_a$ is nondirectional, the critical $z = \pm 2.58$ $[z(1 - \alpha)/2]$.
   e. Confidence interval:

   $$CI_\mu = \overline{X} \pm z_{(1-\alpha)/2}\, \sigma/\sqrt{n}.$$

2. The one-sample situation where the sample is randomly selected from the population and $\sigma$ *is unknown:*

   a. The null hypothesis is stated such that the population mean ($\mu$) is equal to some hypothesized or observed value ($\mu_H$). Therefore, $H_0$: $\mu = \mu_H$.
   b. The alternative may be, depending on the research hypothesis, one of three possibilities:
   $H_a$: $\mu \neq \mu_H$ (nondirectional),
   $H_a$: $\mu > \mu_H$ (directional, greater than), or
   $H_a$: $\mu < \mu_H$ (directional, less than).
   c. The test statistic:

   $$t = \frac{\overline{X} - \mu}{s/\sqrt{n}} \text{ (the one-sample t test statistic).}$$

d. $H_0$: Rejection regions:

When $H_a$ is nondirectional, the rejection region is defined as

$$t_{\alpha/2},\ df = n - 1.$$

When $H_a$ is directional, the rejection region is defined as

$$t_{\alpha},\ df = n - 1.$$

e. Confidence interval:

$$CI_{\mu} = \overline{X} \pm t_{\alpha/2,\ df}\ (s/\sqrt{n}).$$

## FOR REVIEW

1. The hypothesis is a statement about the supposed nature of a population.
2. The null hypothesis ($H_0$) is a statement that denotes no statistical difference between the population parameter and some hypothesized value.
3. The alternative hypothesis ($H_a$) is a statement made in opposition to the null hypothesis such that, if the null hypothesis is rejected, we then fail to reject the alternative hypothesis.
4. The test statistic is a statistical procedure used to make a decision about the value of the population parameter.
5. The null hypothesis rejection region defines the value of the test statistic that indicates the rejection of the null hypothesis. The probability that a calculated test statistic falls in the rejection region is denoted by alpha ($\alpha$).
6. The null hypothesis non-rejection region is the region beneath the middle portion of the sampling distribution curve. Whenever the calculated test statistic falls in this region, the evidence does not permit the rejection of the null hypothesis. The probability that a calculated test statistic will fall in this null hypothesis non-rejection region is denoted by $(1 - \alpha)$.
7. Statistical significance is indicated when a test statistic falls in the null hypothesis rejection region. The level of the statistical significance is indicated by the size of the alpha ($\alpha$) error; the smaller the $\alpha$ error, the greater the significance.
8. The Type I error, or alpha ($\alpha$) error, is the chance we have of rejecting a true null hypothesis. Therefore, when we reject the null hypothesis, at $\alpha = 0.05$ for example, our risk of making a Type I error is $P = 0.05$ or 5%.
9. The Type II error, or beta ($\beta$) error, is the chance we have of accepting a false null hypothesis, or the probability of accepting the null hypothesis when it should have been rejected.

### Chapter Knowledge and Comprehension Application

1. Suppose the average lithium carbonate maintenance level among the population of all BPAD psychiatric patients in a large health-care facility is reported to be 1.3 mEq/L with a standard deviation of 0.3 mEq/L. Suppose

we have a randomly selected sample of 25 patients, with an average lithium carbonate maintenance level of 1.18 mEq/L. Does this information provide evidence that the average maintenance level for the population of psychiatric patients at the facility is significantly different, at $\alpha = 0.05$, than the reported level of 1.3 mEq/L?

2. Suppose the average lithium carbonate maintenance level among the population of all BPAD patients in a large health-care facility is reported to be 1.3 mEq/L. Suppose we have a randomly selected sample of 16 patients from the closed unit, with an average lithium carbonate maintenance level of 1.55 mEq/L and a standard deviation of 0.4 mEq/L. Does this information provide evidence that the average maintenance level for the population of all BPAD psychiatric patients at the facility is greater than 1.3 mEq/L at $\alpha = 0.01$?

**Answers**

1. **Solution:**

***Step 1 Forming the hypotheses:***

$$H_0: \mu = 1.3 \text{ mEq/L}$$

$$H_a: \mu \neq 1.3 \text{ mEq/L}$$

(a two-sided, nondirectional, different-than alternative hypothesis)

***Step 2 Selecting the test statistic:***

$$\begin{aligned} z &= \frac{\overline{X} - \mu}{\sigma/\sqrt{n}}, \\ &= \frac{1.18 \text{ mEq/L} - 1.3 \text{ mEq/L}}{0.3 \text{ mEq/L} \div \sqrt{25}}, \\ &= \frac{-0.12 \text{ mEq/L}}{0.06 \text{ mEq/L}}, \\ z &= -2.00. \end{aligned}$$

***Step 3 Determining the null hypothesis rejection region:***

In Figure 10.4 we can see that the calculated z value (−2.00) falls in the lower rejection region for the nondirectional alternative ($\alpha/2 = 0.025$).

***Step 4 Making a conclusion:***

Since the calculated z value (−2.00) falls in the lower null hypothesis rejection region, we can reject the null hypothesis. We can also conclude, with 95% confidence, that the average lithium carbonate maintenance level for the population of all BPAD patients at the facility is different than 1.3 mEq/L, as reported.

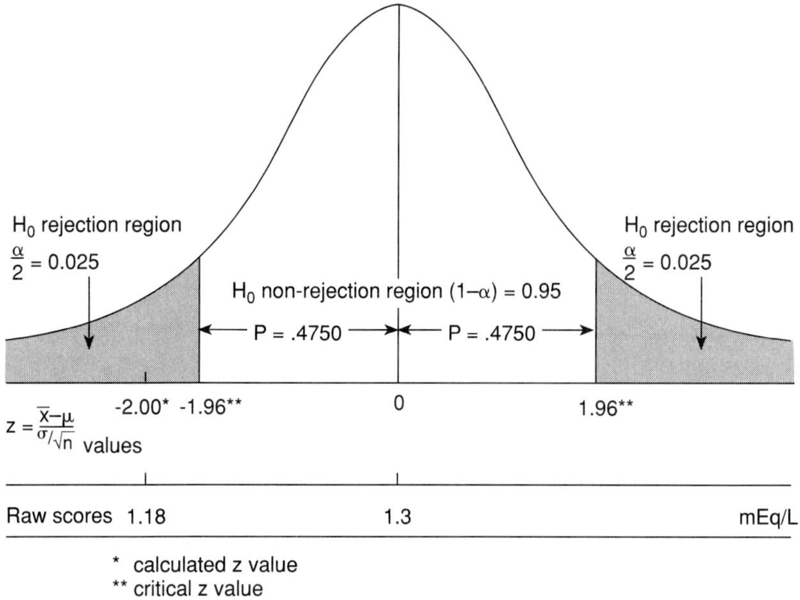


**FIGURE 10.4**
The critical z values (±1.96) associated with $\alpha = 0.05$ and a nondirectional, different-than alternative hypothesis.

**2. Solution:**

### *Step 1 Forming the hypothesis:*

$H_0$: $\mu = 1.3$ mEq/L,

$H_a$: $\mu > 1.3$ mEq/L (a directional, greater-than, alternative hypothesis).

### *Step 2 Selecting the test statistic:*

$$t = \frac{\overline{X} - \mu}{s/\sqrt{n}},$$

(Since in this case σ is unknown, use the t statistic.)

$$= \frac{1.55 \text{ mEq/L} - 1.3 \text{ mEq/L}}{0.4 \text{ mEq/L} \div \sqrt{16}},$$

$$= \frac{0.25 \text{ mEq/L}}{0.1 \text{ mEq/L}},$$

$$t = 2.50.$$

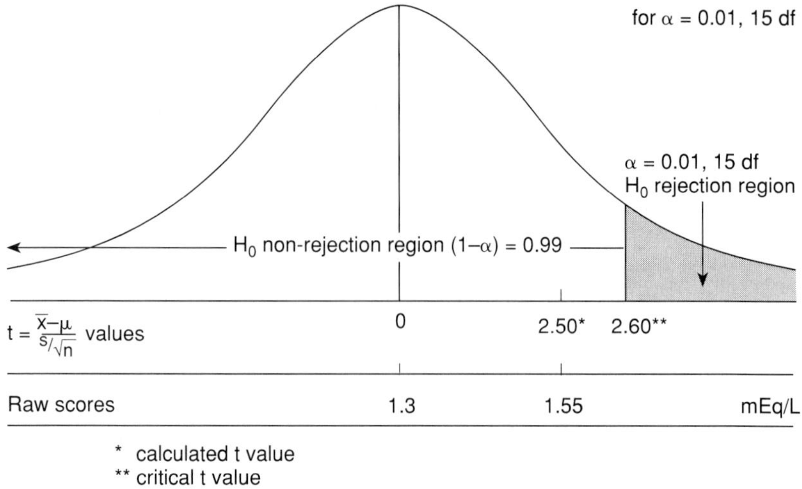


**FIGURE 10.5**
The critical t value (2.60) associated with $\alpha = 0.01$, 15 degrees of freedom, and a one-directional, greater-than test of the null hypothesis.

### *Step 3 Determining the null hypothesis rejection region:*

In Figure 10.5 we can see that the calculated t value (2.50) does not fall in the rejection region ($\alpha = 0.01$, 15 df).

### *Step 4 Making a conclusion:*

Since the calculated t value (2.50) does not fall in the null hypothesis rejection region, we fail to reject the null hypothesis. We conclude that the sample of 16 BPAD patients does not provide sufficient evidence to indicate that the lithium carbonate maintenance level for the population is greater than 1.3 mEq/L at $\alpha = 0.01$.

CHAPTER

**11**

# Two-sample Tests of Hypotheses About the Difference Between Two Population Means

*"Factual evidence can never prove a hypothesis; it can only fail to disprove it, which is what we generally mean when we say, somewhat inexactly, that the hypothesis is confirmed. . . ."*

MILTON FRIEDMAN

In the previous chapter, we used single samples to make one-sample tests of hypotheses about the means of populations. In this chapter, the hypothesis tests described will be based on two samples, and we will follow the same basic steps of hypothesis testing that we followed when using one sample. These four steps are:

Step 1 Forming the hypotheses

Step 2 Selecting the test statistic

Step 3 Determining the null hypothesis rejection region, and

Step 4 Making a conclusion

Before we begin our discussion of two-sample tests of hypotheses, it is necessary for us to differentiate between independent and dependent samples. Our eventual choice of a test statistic will be determined, in part, by the independent or the dependent nature of the samples used.

## INDEPENDENT AND DEPENDENT SAMPLES

In Chapter 10 we dealt with a single sample obtained randomly from a single population. Then, our inference about the mean of a population was based on a single sample. In this chapter we will make inferences about the means of two populations. This requires the selection of two samples, one from each population. The methods used to select samples determines whether the two samples are independent or dependent.

### Independent Sampling Procedures

Two samples are **independent** if the cases selected from one population are unrelated to and mutually exclusive from the cases selected from the second population. An example of the sampling procedure that would result in an independent sample follows.

A hospital administrator is interested in comparing the starting salaries of male, newly graduated, registered nurses with the starting salaries of their female counterparts. To ensure two independent samples, the administrator could randomly select two separate samples. One random sample of 200 female, newly graduated, registered nurses could be selected from all the female nurses graduating this year. The other, a separate random sample of 200 male, newly graduated, registered nurses, could be selected from all the male nurses graduating this year. The average starting salaries for each sample could then be compared.

This sampling procedure results in independence. The selection of the female nursing students occurs separately and exclusively from the selection of male nursing students. Two different, unrelated random samples are selected, each from a different population.

### Dependent Sampling Procedures

Two samples are **dependent** if the cases selected from one population somehow relate to the cases selected from the other sample. This relatedness can result from several different sampling procedures. For example, the cases from each of the populations can be matched to form **matched pairs** of similar cases. Using the previous situation, the hospital administrator could decide to match the male and female registered nurses according to some other variable, such as grade point average upon graduation. After matching each of the newly graduated female nurses with a newly graduated male nurse having approximately the same grade point average, 200 of the matched pairs could be randomly selected and their starting salaries compared.

This method of sampling, in which the cases are paired by grade point average before the random selection of pairs occurs, results in dependency.

For each newly graduated, female, registered nurse selected, a male registered nurse (with a similar grade point average) is selected. Other pairing methods are commonly used to create dependent samples. **Genetic pairing** occurs when sets of identical twins or litter mates serve as dependent pairs. Also, self-pairing results in dependency. **Self-pairing** occurs when subjects from "before-and-after" studies serve as their own controls. That is, in a study to determine the effect of a diet treatment, the subjects' weights are measured before and after the diet treatment. Thus, each subject's before-diet and after-diet weights are compared. Whenever the samples selected are somehow related, as in pairing, the samples are dependent in nature.

## THE TWO-SAMPLE TEST OF A HYPOTHESIS ABOUT THE MEANS OF TWO POPULATIONS WHEN $\sigma$ IS KNOWN

Suppose the average lithium carbonate maintenance level among the population of all BPAD psychiatric patients in a large health-care facility is reported to be 1.3 mEq/L. A health-care researcher is interested in comparing the lithium maintenance levels of the population of all male BPAD psychiatric patients with the lithium maintenance levels of the population of all female BPAD psychiatric patients. Suppose a relatively large sample of 49 patients is randomly selected from each of the populations. Do the average lithium maintenance levels of 1.25 mEq/L for the female sample and 1.35 mEq/L for the male sample provide evidence (at $\alpha = 0.05$) that the lithium maintenance levels *differ* according to gender? Since $n_1$ and $n_2$ are large (49), assume that $\sigma$ (0.3 mEq/L) is equal to $\sigma_1$ (the population standard deviation for the women) and $\sigma_2$ (the population standard deviation for the men).

### *Step 1 Forming the hypotheses:*

$$H_0: \mu_1 = \mu_2,$$

where $\mu_1$ represents the average lithium maintenance level of the population of female BPAD patients and $\mu_2$ represents the average lithium maintenance level of the population of male BPAD patients; and

$$H_a: \mu_1 \neq \mu_2,$$

the nondirectional, two-sided test of difference is indicated by the word "differ" in the example.

### *Step 2 Selecting the test statistic:*

Let us review what information we have been given in this example.

*The female random sample:*

$n_1 = 49$ BPAD patients' lithium maintenance levels,

$\overline{X}_1 = 1.25$ mEq/L (the average of the 49 lithium maintenance levels),

$\sigma_1 = \sigma = 0.3$ mEq/L.

*The male random sample:*

$n_2 = 49$ BPAD patients' lithium maintenance levels,

$\overline{X}_2 = 1.35$ mEq/L (the average of the 49 lithium maintenance levels),

$\sigma_2 = \sigma = 0.3$ mEq/L.

We assume that the standard deviations for the population of women and the population of men ($\sigma_1$ and $\sigma_2$, respectively) are equal to the standard deviation of the population of men and women ($\sigma = 0.3$ mEq/L). Thus, we can use a z test statistic designed to test hypotheses about the means of two populations. In this situation the test statistic takes the form of

$$z = \frac{(\overline{X}_1 - \overline{X}_2)}{\sigma\sqrt{\dfrac{1}{n_1} + \dfrac{1}{n_2}}},$$

where $\overline{X}_1$ and $\overline{X}_2$ are estimates of the true mean of the respective populations $\mu_1$ and $\mu_2$, and

$$\sigma\sqrt{\frac{1}{n_1} + \frac{1}{n_2}}$$

is the standard error. The properties of the sampling distribution we used with the one-sample z test can be applied here to the two-sample z test. If both $\overline{X}_1$ and $\overline{X}_2$ are normally distributed, then it follows that the difference $(\overline{X}_1 - \overline{X}_2)$ is also normally distributed. Suppose the distribution of $\overline{X}_1$ has a true mean $\mu_1$ and a standard error $\sigma_1/\sqrt{n_1}$, and the distribution of $\overline{X}_2$ has a true mean $\mu_2$ and a standard error $\sigma_2/\sqrt{n_2}$. Therefore, the distribution of the difference between $\overline{X}_1$ and $\overline{X}_2$ $(\overline{X}_1 - \overline{X}_2)$ has a true mean $(\mu_1 - \mu_2)$ and a standard error of

$$\sqrt{\frac{\sigma_1^2}{n_1} + \frac{\sigma_2^2}{n_2}}.$$

When we assume $\sigma_1 = \sigma_2 = \sigma$, then the standard error becomes

$$\sqrt{\frac{\sigma^2}{n_1} + \frac{\sigma^2}{n_2}}, \quad \text{or} \quad \sigma\sqrt{\frac{1}{n_1} + \frac{1}{n_2}}.$$

In this case,

$$
\begin{aligned}
z &= \frac{\overline{X}_1 - \overline{X}_2}{\sigma\sqrt{\frac{1}{n_1} + \frac{1}{n_2}}}, \\
&= \frac{1.25 \text{ mEq/L} - 1.35 \text{ mEq/L}}{0.3 \text{ mEq/L}/\sqrt{\frac{1}{49} + \frac{1}{49}}}, \\
&= \frac{-0.1 \text{ mEq/L}}{0.3 \text{ mEq/L} \sqrt{.041}}, \\
&= \frac{-0.1 \text{ mEq/L}}{.06 \text{ mEq/L}}, \\
z &= -1.67.
\end{aligned}
$$

The two-sample z test is indicated here because we assume $\sigma_1 = \sigma_2 = \sigma =$ 0.3 mEq/L and we have two, independent random samples ($\overline{X}_1$ and $\overline{X}_2$) that represent two independent population means ($\mu_1$ and $\mu_2$).

### *Step 3 Determining the null hypothesis rejection region:*

Since we have a nondirectional, two-sided alternative hypothesis of difference with $\alpha = 0.05$, our critical z values ($z = \pm 1.96$, from Appendix Table A) are indicated in Figure 11.1.

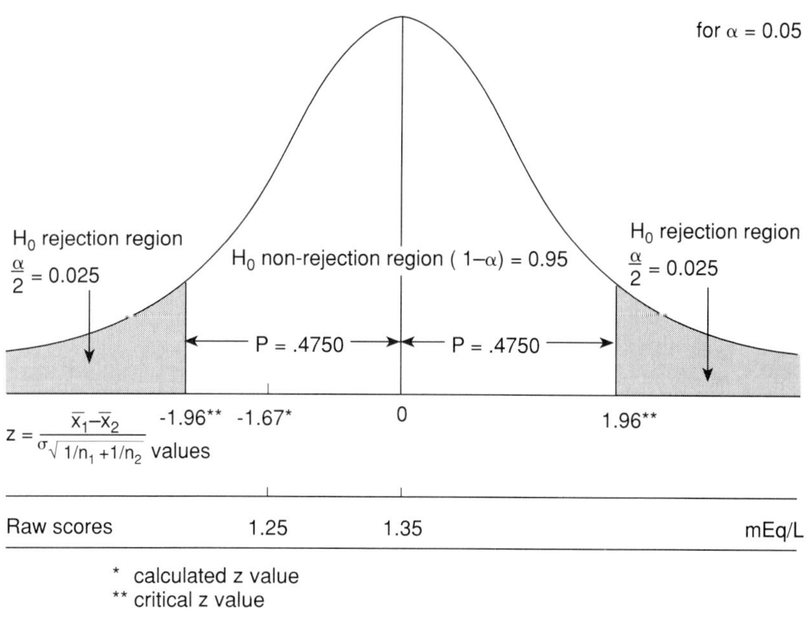


**FIGURE 11.1**
The critical z values (±1.96) associated with $\alpha/2 = 0.025$ and a nondirectional, two-sided test of the null hypothesis.

*Step 4 Making a conclusion:*

Notice that the calculated z value of $-1.67$ does not fall in the null hypothesis rejection region (area outside $z = \pm1.96$). Therefore, we fail to reject the null hypothesis. We can conclude that the sample averages ($\overline{X}_1 = 1.25$ mEq/L and $\overline{X}_2 = 1.35$ mEq/L) are not different enough for us to be 95% confident that the average lithium carbonate maintenance level for the population of female BPAD patients is different than that of the population of male BPAD patients.

As with the one-sample situation we could retest, after selecting larger samples, or we could create a 95% confidence interval around the true mean difference such that

$$95\%CI_{(\mu_1-\mu_2)} = (\overline{X}_1 - \overline{X}_2) \pm z_{(.4750)}\sigma\sqrt{\frac{1}{n_1} + \frac{1}{n_2}},$$

$$= (1.25 \text{ mEq/L} - 1.35 \text{ mEq/L}) \pm 1.96\left(0.3 \text{ mEq/L}\sqrt{\frac{1}{49} + \frac{1}{49}}\right),$$

$$= (-0.1 \text{ mEq/L}) \pm 1.96\ (.06 \text{ mEq/L}),$$

$$95\%CII_{(\mu_1-\mu_2)} = [-.218 \text{ mEq/L to } .018 \text{ mEq/L}].$$

Therefore, we are 95% confident that the interval [−.218 mEq/L to mEq/L .018] includes the true mean difference of lithium carbonate for the female and male BPAD patients.

## THE TWO-SAMPLE TEST OF A HYPOTHESIS ABOUT THE MEANS OF TWO POPULATIONS WHEN σ IS UNKNOWN

In Chapter 10, the t distribution was used when the standard deviation of the population ($\sigma$) was unknown. Here, we will discuss two additional t test statistics that can be used to test the difference between groups when $\sigma$ is unknown. One, the **pooled t test,** is used with independent samples; the other, the **paired t test,** is used with dependent samples.

### The Pooled t Test for Independent Samples

Suppose the average lithium carbonate maintenance level for a random sample of nine, closed-unit, BPAD psychiatric patients is 1.4 mEq/L with a standard deviation (s) of 0.20 mEq/L. Also, the average lithium carbonate maintenance level for a random sample of 16 open-unit patients is 1.0 mEq/L with a standard deviation (s) of 0.4 mEq/L. Do these sample statistics provide sufficient evidence at $\alpha = 0.01$ to claim that the average lithium carbonate maintenance level for the population of closed-unit, BPAD psychiatric patients is greater than the average lithium carbonate maintenance level for the population of open-unit, BPAD psychiatric patients?

***Step 1 Forming the hypotheses:***

$$H_0: \mu_1 = \mu_2,$$

where $\mu_1$ represents the average lithium maintenance level of the population of closed-unit BPAD patients, and $\mu_2$ represents the average lithium maintenance level of the population of open-unit BPAD patients, and

$$H_a: \mu_1 > \mu_2,$$

a directional, one-sided, greater-than alternative hypothesis is indicated.

***Step 2 Selecting the test statistic:***

Before we continue, let us review the information already given for this example.
*The closed-unit random sample:*

$n_1$ = nine randomly selected BPAD patients' lithium maintenance levels,

$\overline{X}_1$ = 1.4 mEq/L (the average of the sample of nine lithium maintenance levels),

$s_1$ = 0.2 mEq/L (the standard deviation of the sample of nine lithium maintenance levels).

*The open-unit random sample:*

$n_2$ = 16 randomly selected BPAD patients' lithium maintenance levels,

$\overline{X}_2$ = 1.0 mEq/L (the average of the sample of 16 lithium maintenance levels),

$s_2$ = 0.4 mEq/L (the standard deviation of the sample of 16 lithium maintenance levels).

The population standard deviation $\sigma_1$ and $\sigma_2$ are unknown. The closed-unit population was randomly sampled independently of the open-unit population. Also, at the time the samples were selected, the two populations were mutually exclusive. Therefore, we have independent, relatively small samples taken from exclusive populations with unknown population standard deviations. In this situation, the pooled t-test statistic is indicated and may be written as

$$t = \frac{(\overline{X}_1 - \overline{X}_2)}{s_p\sqrt{\frac{1}{n_1} + \frac{1}{n_2}}}.$$

This test statistic is comparable to the previously discussed, two-sample z test, with one exception. Previously, we were able to assume that $\sigma_1 = \sigma_2 = \sigma$. Here, since we have two small samples of unequal size, we use $s_1$ and $s_2$ together, with their respective degrees of freedom, to calculate the pooled estimate of the standard deviation ($s_p$). The pooled standard deviation ($s_p$) is found as

$$s_p = \sqrt{\frac{s_1^2\,(n_1 - 1) + s_2^2\,(n_2 - 1)}{(n_1 - 1) + (n_2 - 1)}}.$$

Under the square root sign in the numerator, the variance of the closed-unit sample ($s_1^2$) is multiplied by the degrees of freedom of the closed-unit sample ($n_1 - 1$). Likewise, the variance of the open-unit sample ($s_2^2$) is multiplied by the degrees of freedom of the open-unit sample ($n_2 - 1$). Below the square root sign in the denominator are the total degrees of freedom for both samples combined, $(n_1 - 1) + (n_2 - 1)$. This is sometimes written as $(n_1 + n_2) - 2$. For this pooled t test statistic, degrees of freedom equal $n_1 + n_2 - 2$.

First, let us calculate the pooled standard deviation which we'll use later when we calculate the pooled standard error

$$s_p \sqrt{\frac{1}{n_1} + \frac{1}{n_2}}.$$

Using the sample information from the previous example,

$$
\begin{aligned}
s_p &= \sqrt{\frac{s_1^2\,(n - 1) + s_2^2\,(n - 2)}{(n_1 - 1) + (n_2 - 1)}},\\
&= \sqrt{\frac{(0.20)^2 \cdot (9 - 1) + (0.40)^2 \cdot (16 - 1)}{(9 - 1) + (16 - 1)}},\\
&= \sqrt{\frac{(.32) + (2.40)}{23}},\\
&= \sqrt{\frac{2.72}{23}},\\
&= \sqrt{.118},\\
S_p &= 0.34.
\end{aligned}
$$

Now we can calculate the test statistic using $s_p$ (0.34 mEq/L) in the calculation of the estimate of the pooled standard error.

$$
\begin{aligned}
t &= \frac{\overline{X}_1 - \overline{X}_2}{s_p\sqrt{\frac{1}{n_1} + \frac{1}{n_2}}},\\
&= \frac{1.4 - 1.0}{.34\sqrt{\frac{1}{9} + \frac{1}{16}}},\\
&= \frac{.4}{.34\sqrt{.111 + .063}},\\
&= \frac{.4}{.34\sqrt{.173}},\\
&= \frac{.4}{.34\,(.416)},\\
t &= 2.84.
\end{aligned}
$$

The calculated t value is equal to 2.84.

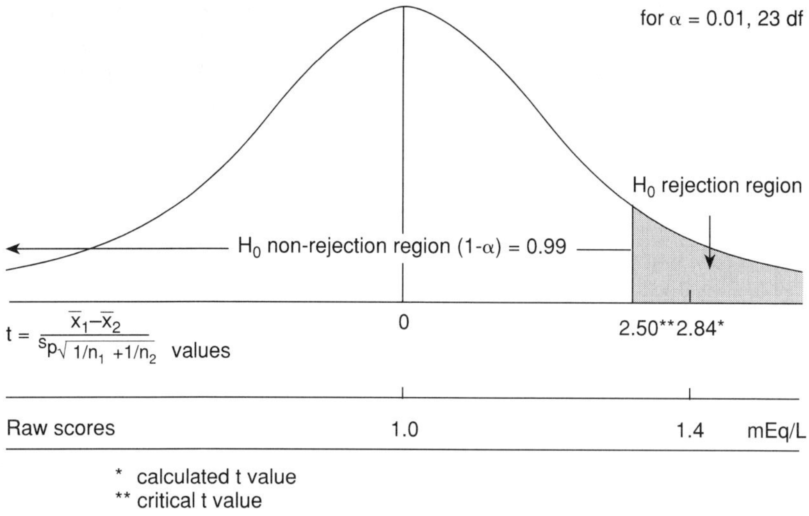


**FIGURE 11.2**
The critical t value (2.50) associated with $\alpha = 0.01$, 15 degrees of freedom, and a directional, greater-than test of the null hypothesis.

### *Step 3 Determining the null hypothesis rejection region:*

Since the alternate hypothesis indicates a directional, greater-than, one-sided test with $\alpha = 0.01$, our critical t value (with $n_1 + n_2 - 2 = 23$ df) is $t = 2.50$. This critical t value and the $\alpha = 0.01$ rejection region are shown in Figure 11.2.

The calculated t value of $t = 2.84$ is greater than the critical t value of $t = 2.50$ ($\alpha = 0.01$, 23 df) and falls in the greater-than null hypothesis rejection region.

### *Step 4 Making a conclusion:*

Reject the null hypothesis at $\alpha = 0.01$, 23 df and conclude that the closed-unit BPAD patients are maintained with significantly greater average lithium carbonate levels than the open-unit BPAD patients.

A 99% confidence interval of the true mean difference in lithium carbonate maintenance levels for the closed- and open-unit patients is given by

$$
\begin{aligned}
99\%CI_{(\mu_1-\mu_2)} &= (\overline{X}_1 - \overline{X}_2) \pm t_{.005,\ 23\ df}\, s_p\sqrt{\frac{1}{n_1} + \frac{1}{n_2}},\\
&= (1.4\ \text{mEq/L} - 1.0\ \text{mEq/L}) \pm 2.80\left(.34\ \text{mEq/L}\sqrt{\frac{1}{9} + \frac{1}{16}}\right),\\
&= (.40\ \text{mEq/L}) \pm 2.80\ (.34\ \text{mEq/L} \times .416),\\
&= (.40\ \text{mEq/L}) \pm 2.80\ (.14\ \text{mEq/L}),\\
&= (.40\ \text{mEq/L}) \pm .392\ \text{mEq/L},\\
99\%CI_{(\mu_1-\mu_2)} &= [.008\ \text{mEq/L to } .792\ \text{mEq/L}].
\end{aligned}
$$

**TABLE 11.1** Subjective Estimates of Anxiety Made by Seven BPAD Psychiatric Outpatients Before and After Lorazepam Treatment

| | Subjective Estimates of Anxiety | |
|---|---|---|
| *Patient* | *Before Treatment* | *After Treatment* |
| 1 | 06 | 06 |
| 2 | 08 | 06 |
| 3 | 09 | 09 |
| 4 | 10 | 07 |
| 5 | 07 | 05 |
| 6 | 08 | 05 |
| 7 | 09 | 08 |

Therefore, we are 99% confident that the interval [.008 mEq/L to .792 mEq/L] includes the true difference in average lithium maintenance for the closed- and open-unit patients.

## The Paired t Test for Dependent Samples

Suppose a randomly selected sample of seven BPAD psychiatric patients, treated as outpatients, volunteer to take part in a study. This study will determine the effect a minimum therapeutic dose of the antianxiety medication lorazepam (2 mg.) has on their subjective estimates of anxiety. Each patient responds to a subjective anxiety inventory questionnaire, both before and one hour after receiving the minimum therapeutic dose of lorazepam (2 mg.). The resulting data are presented in Table 11.1.

Do these data provide evidence to support the claim that the subjective estimates of anxiety among these seven BPAD psychiatric outpatients were significantly lower *after* treatment than before treatment? Test at $\alpha = 0.01$.

Before we continue with the solution, it is important to point out that this example represents a commonly used method of pairing called self-pairing. As described earlier in this chapter, in self-pairing, each case (patient) serves as his or her own control. Here, each of the outpatients subjectively estimates his or her before-treatment anxiety level. That anxiety level is paired with the same subjects' after-treatment subjective estimate of anxiety.

The null hypothesis, for the paired test, is that the average difference among the pairs is equal to zero. In this example, the null hypothesis is that the average difference between the paired (before and after treatment) subjective anxiety estimates, for each of the seven patients combined, equals zero. In effect, we consider the seven differences—generated from the paired anxiety scores for each of the patients—as a single sample. Thus, we sum the seven differences and divide the total difference by seven to find the average difference. This average difference is symbolized as $\bar{d}$. Therefore, $\bar{d}$ is the best estimate of the true mean of the population of differences, $\mu_D$.

***Step 1 Forming the hypotheses:***

$$H_0: \mu_D = 0,$$

where $\mu_D$ represents the true mean of the population of differences such that $\mu_1 - \mu_2 = \mu_D$.

The null hypothesis in this case is that the lorazepam minimal dose (2 mg.) does not affect the subjective anxiety estimates of BPAD psychiatric patients treated on an outpatient basis. In other words, there is no difference in the before and after anxiety estimates for patients receiving the lorazepam treatment described.

$$H_a: \mu_D < 0,$$

where $\mu_D$ represents the true mean of the population of differences such that $\mu_1 - \mu_2 = \mu_D$.

The alternative hypothesis is that the anxiety estimates *after* the treatment were significantly lower than were the anxiety estimates *before* treatment among the BPAD psychiatric outpatients.

***Step 2 Selecting the test statistic:***

The paired t test statistic for a dependent sample is

$$t = \frac{\bar{d} - \mu_D}{s_d/\sqrt{n}},$$

where

$\bar{d}$ = the mean difference generated from the sample differences,

$\mu_D$ = the true mean difference of the population of differences,

$s_d$ = the standard deviation of the sample differences such that

$$s_d = \sqrt{\frac{\Sigma d^2 - \left[\frac{(\Sigma d)^2}{n}\right]}{n - 1}}, \text{ and}$$

n = the number of pairs in the sample.

The paired t test statistic is, after all, just a one-sample, t test statistic used on the sample of differences between the pairs. Thus, for the one-sample t test statistic, $df = n-1$, where n is the number of pairs.

We used $\bar{d}$ as the estimate of $\mu_D$. When the null hypothesis is such that $\mu_D = 0$, then the test statistic

$$t = \frac{\bar{d} - \mu_D}{s_d/\sqrt{n}} \quad \text{becomes} \quad t = \frac{\bar{d}}{s_d/\sqrt{n}}.$$

Using the data presented in Table 11.2, we will now calculate the paired t test statistic.

**TABLE 11.2** Paired t Test Statistic Calculated from Subjective Estimates of Anxiety

| Patient | Anxiety Before Treatment | Anxiety After Treatment | Difference (d) | Squared Difference ($d^2$) |
|---|---|---|---|---|
| 1 | 06 | 06 | 0 | 0 |
| 2 | 08 | 06 | −2 | 4 |
| 3 | 09 | 09 | 0 | 0 |
| 4 | 10 | 07 | −3 | 9 |
| 5 | 07 | 05 | −2 | 4 |
| 6 | 08 | 05 | −3 | 9 |
| 7 | 09 | 08 | −1 | 1 |
| n = 7 | | | $\sum d = -11$ | $\sum d^2 = 27$ |

Difference column values are generated as ''after'' minus ''before'' differences. This order corresponds to the alternative (less-than) hypothesis. Once an order of subtraction has been decided upon, all subtractions between all the pairs must be consistent with that order. The negative differences indicate only a decrease in anxiety levels from the ''before'' to the ''after'' condition. The average difference between the pairs is

$$\bar{d} = \frac{\Sigma d}{n} = \frac{-11}{7} = -1.57.$$

The standard deviation of the difference between the pairs is

$$s_d = \sqrt{\frac{\Sigma d^2 - \left[\frac{(\Sigma d)^2}{n}\right]}{n - 1}},$$

$$= \sqrt{\frac{27 - \left[\frac{(-11)^2}{7}\right]}{7 - 1}},$$

$$= \sqrt{\frac{27 - \left[\frac{121}{7}\right]}{6}},$$

$$= \sqrt{\frac{27 - 17.29}{6}},$$

$$= \sqrt{1.62},$$

$$s_d = 1.27.$$

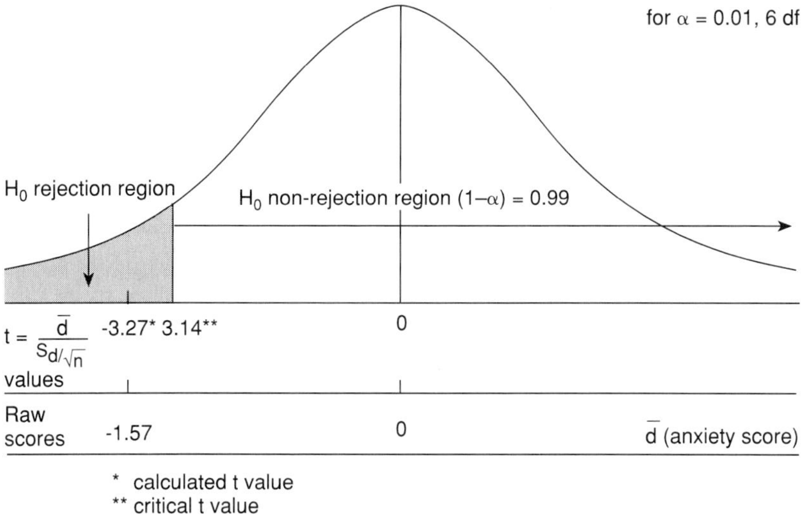


**FIGURE 11.3**
Critical t value (−3.14) associated with $\alpha = 0.01$, 6 degrees of freedom, and a directional, less-than test of the null hypothesis.

Therefore, the paired t test statistic is

$$t = \frac{\bar{d}}{s_d/\sqrt{n}}$$
$$= \frac{-1.57}{1.27/\sqrt{7}}$$
$$= \frac{-1.57}{0.48}$$
$$t = -3.27.$$

### *Step 3 Determining the null hypothesis rejection region:*

Since a directional, one-sided, less-than, alternative hypothesis is indicated by the research question, with $\alpha = 0.01$, our critical t value (with $n - 1 = 6$ df) is $t = -3.14$. This critical t value and the $\alpha = 0.01$ rejection region are shown in Figure 11.3.

### *Step 4 Making a conclusion:*

The calculated t value ($t = -3.27$, Figure 11.3) falls within the null hypothesis rejection region defined by the critical t value −3.14. Therefore, we reject the null hypothesis. We can conclude that the subjective anxiety estimates made one hour after receiving a minimal lorazepam dose (2 mg.) are significantly lower than those estimates made before receipt of treatment at $\alpha = 0.01$ with 6 df.

We can create a 99% confidence interval on the true mean difference ($\mu_D$) as

$$99\%CI_{\mu_D} = \bar{d} \pm t_{.005,6\ df,} (s_d/\sqrt{n}),$$
$$= -1.57 \pm 3.71\ (1.27/\sqrt{7}),$$
$$= 1.57 \pm 3.71\ (0.48),$$
$$= -1.57 \pm (1.78),$$
$$99\%CI_{\mu_D} = [-3.35 \text{ to } 0.21].$$

This confidence interval may be interpreted as indicating the interval that includes the true mean of the differences in subjective anxiety estimates (before and after the minimum lorazepam maintenance dose of 2 mg.) among psychiatric outpatients for 99 such samples out of 100.

## SUMMARY

In this chapter we have considered two-sample tests of hypotheses about the means of populations. These tests are summarized here:

1. For the two-sample situation when $\sigma$ *is known,*
   a. The null hypothesis states that the means of the two populations are equal.

   $$H_0: \mu_1 = \mu_2.$$

   b. The alternative hypothesis may be one of three possibilities:

   $H_a: \mu_1 \neq \mu_2$ (nondirectional, different-than alternative),

   $H_a: \mu_1 > \mu_2$ (directional, greater-than alternative), or

   $H_a: \mu_1 < \mu_2$ (directional, less-than alternative).

   c. The test statistic:

   $$z = \frac{\bar{X}_1 - \bar{X}_2}{\sigma\sqrt{\frac{1}{n_1} + \frac{1}{n_2}}} \quad \text{(the two-sample z-test statistic.)}$$

   d. $H_0$: rejection regions:
   For $\alpha = 0.05$, when $H_a$ is directional, the critical $z = \pm 1.645$,
   For $\alpha = 0.05$, when $H_a$ is nondirectional, the critical $z = \pm 1.96$ $[z_{(1-\alpha)/2}]$,
   For $\alpha = 0.01$, when $H_a$ is directional, the critical $z = \pm 2.33$,
   For $\alpha = 0.01$, when $H_a$ is nondirectional, the critical $z = \pm 2.58$ $[z_{(1-\alpha)/2}]$.

   e. Confidence interval:

   $$CI_{(\mu_1 - \mu_2)} = (\bar{X}_1 - \bar{X}_2) \pm z_{(1-\alpha)/2}\left(\sigma\sqrt{\frac{1}{n_1} + \frac{1}{n_2}}\right).$$

2. For the two-sample situation when $\sigma$ *is unknown* and the samples are independently selected,
   a. The null hypothesis states that the means of the two populations are equal:

$$H_0: \mu_1 = \mu_2.$$

   b. The alternative may be, depending on the research hypothesis, one of these possibilities:

$H_a: \mu_1 \neq \mu_2$ (nondirectional, different-than alternative),

$H_a: \mu_1 > \mu_2$ (directional, greater-than alternative), or

$H_a: \mu_1 < \mu_2$ (directional, less-than alternative).

   c. The test statistic:

$$t = \frac{\overline{X}_1 - \overline{X}_2}{s_p\sqrt{\frac{1}{n_1} + \frac{1}{n_2}}} \quad \text{(the pooled t test),}$$

   where $s_p$ is the pooled, standard deviation of the sample

$$s_p = \sqrt{\frac{[s_1^2\,(n_1 - 1)] + [s_2^2\,(n_2 - 1)]}{n_1 + n_2 - 1}}.$$

   d. $H_0$: Rejection regions:
   When $H_a$ is nondirectional, the rejection region is defined as

$$t_{\alpha/2},\ df = (n_1 - 1) + (n_2 - 1) \text{ or}$$
$$t_{\alpha/2},\ df = (n_1 + n_2) - 2.$$

   When $H_a$ is directional, the rejection region is defined as

$$t_{\alpha},\ df = (n_1 - 1) + (n_2 - 1) \text{ or } t_{\alpha},\ df = (n_1 + n_2) - 2.$$

   e. Confidence interval:

$$CI_{(\mu_1 - \mu_2)} = (\overline{X}_1 - \overline{X}_2) \pm t_{\alpha/2}\left(s_p\sqrt{\frac{1}{n_1} + \frac{1}{n_2}}\right), \text{ and } df = (n_1 + n_2) - 2.$$

3. For the paired t test for dependent samples when $\sigma$ *is unknown,*
   a. The null hypothesis states that the true mean of the population of differences ($\mu_D$) is equal to zero:

   $H_0: \mu_D = 0$, where $\mu_D$ represents the true mean of the population of differences such that $\mu_1 - \mu_2 = \mu_D$.

   b. The alternative hypothesis may be one of these possibilities:

$H_a: \mu_D \neq 0$ (nondirectional, different-than alternative),

$H_a: \mu_D > 0$ (directional, greater-than alternative), or

$H_a: \mu_D < 0$ (directional, less-than alternative).

c. The test statistic

$$t = \frac{\bar{d}}{s_d/\sqrt{n}} \qquad \text{where}$$

$\bar{d}$ (the average difference between the sampled pairs) $= \frac{\Sigma d}{n}$ and,

$s_d$ (the standard deviation of the differences between the sampled pairs) $= \sqrt{\frac{\Sigma d^2 - \left[\frac{(\Sigma d)^2}{n}\right]}{n - 1}}$.

d. $H_0$: Rejection regions:
When $H_a$ is nondirectional, the rejection region is defined as:

$t_{\alpha/2}$, df $= n - 1$ (where n is equal to the number of pairs).

When $H_a$ is directional, the rejection region is defined as:

$t_{\alpha}$, df $= n - 1$ (where n is equal to the number of pairs).

e. Confidence interval:

$$CI_{\mu_D} = \bar{d} \pm t_{\alpha/2}(s_d/\sqrt{n}), \text{ and}$$

$$df = n - 1 \text{ (where n is the number of pairs).}$$

## FOR REVIEW

1. Independent samples are selected in a way that will ensure they are unrelated and mutually exclusive.
2. Dependent samples are selected in a way that will ensure some degree of relatedness. Dependent samples may be formed of mutually exclusive, yet dependent pairs (as with pairs of identical twins), or with matched pairs. Dependent samples may be formed of pairs of measurements taken from the same case as with self-pairing (when the case serves as its own control).

### Chapter Knowledge and Comprehension Application

1. Suppose a randomly selected sample of eight, BPAD psychiatric patients, consisting of four men and four women, is chosen from among the population of outpatients at a large health-care facility. They will participate in a study designed to determine the effect of a minimum therapeutic dose

of the antianxiety medication lorazepam on the patients' subjective estimates of anxiety levels. Each patient responds to a subjective anxiety inventory questionnaire, both before and one hour after receiving the minimum therapeutic dose of lorazepam. The data resulting from the study are presented in Table 11.3.

**TABLE 11.3** Subjective Estimates of Anxiety Made by Eight BPAD Psychiatric Outpatients Before and After Lorazepam Treatment

| | Subjective Estimates of Anxiety | |
|---|---|---|
| *Patient/Gender* | *Before Treatment* | *After Treatment* |
| 1/M | 08 | 06 |
| 2/M | 08 | 06 |
| 3/M | 09 | 06 |
| 4/M | 10 | 08 |
| 5/F | 10 | 08 |
| 6/F | 07 | 05 |
| 7/F | 09 | 08 |
| 8/F | 08 | 06 |

a. Do these data provide evidence to support the claim that the subjective estimates of anxiety among the eight BPAD psychiatric outpatients were significantly *lower* after lorazepam treatment than before? Test at $\alpha = 0.05$.

b. Do these data provide evidence to support the claim that subjective estimates of anxiety recorded by the male patients *after* treatment were different than the subjective anxiety estimates recorded by the female patients *after* treatment? Test at $\alpha = 0.05$.

**Answers**
**Solution (a):**

***Step 1 Forming the hypotheses:***

$H_0$: $\mu_D = 0$,

$H_a$: $\mu_D = < 0$ (directional, less-than alternative hypothesis).

### *Step 2 Selecting the test statistic:*

$$t = \frac{\bar{d}}{s_d/\sqrt{n}}$$ (where $\bar{d}$ and $s_d$ are calculated using the data from Table 11.2).

| Patient/ Gender | Anxiety Before | Anxiety After | d | $d^2$ |
|---|---|---|---|---|
| 1/M | 08 | 06 | −2 | 4 |
| 2/M | 08 | 06 | −2 | 4 |
| 3/M | 09 | 06 | −3 | 9 |
| 4/M | 10 | 08 | −2 | 4 |
| 5/F | 10 | 08 | −2 | 4 |
| 6/F | 07 | 05 | −2 | 4 |
| 7/F | 09 | 08 | −1 | 1 |
| 8/F | 08 | 06 | −2 | 4 |
| n = 8 | | | Σd = −16 | $\Sigma d^2 = 34$ |

$$\bar{d} = \frac{\Sigma d}{n} = \frac{-16}{8} = -2.0, \text{ and}$$

$$s_d = \sqrt{\frac{\Sigma d^2 - \left[\frac{(\Sigma d)^2}{n}\right]}{n-1}} = \sqrt{\frac{34 - \left[\frac{(16)^2}{8}\right]}{7}}$$

$$= \sqrt{\frac{34 - 32}{7}} = 0.53. \text{ Therefore,}$$

$$t = \frac{-2.0}{0.53\sqrt{8}} = \frac{-2.0}{0.53(2.83)} = \frac{-2.0}{1.50}$$

$$t = -1.33.$$

### *Step 3 Determining the null hypothesis rejection region:*

In Figure 11.4 we can see that the calculated t value (−1.33) does not fall in the $H_0$ rejection region ($\alpha = 0.05$, df = 7).

### *Step 4 Making a conclusion:*

Since the calculated t value did not fall in the null hypothesis rejection region, we fail to reject the null hypothesis. We conclude that the subjective anxiety estimates made by the eight patients after treatment were not significantly lower than the subjective anxiety estimates made before the treatment at $\alpha$ = 0.05 and 7 df.

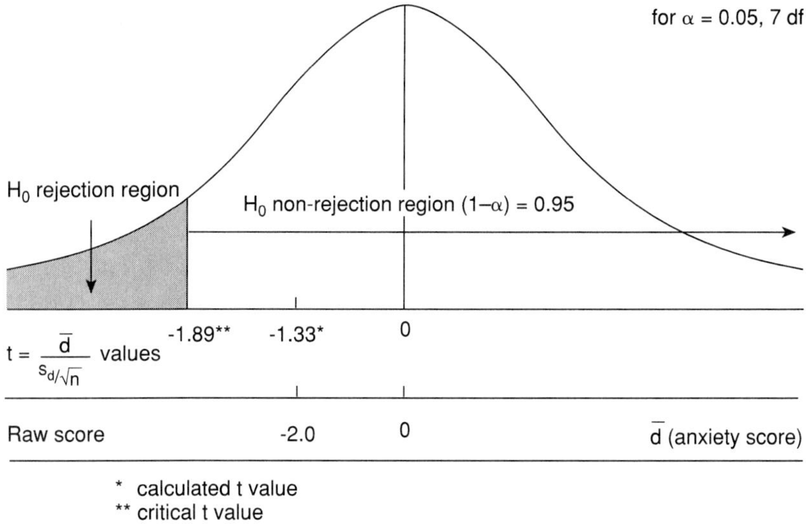


**FIGURE 11.4**
The critical t value (−1.89) associated with $\alpha = 0.05$, 7 degrees of freedom, and a directional, less-than test of the null hypothesis.

## Solution (b):

### *Step 1 Forming the hypotheses:*

$H_0$: $\mu_1 = \mu_2$,

$H_a$: $\mu_1 \neq \mu_2$ (a nondirectional, alternative hypothesis).

### *Step 2 Selecting the test statistic:*

$$t = \frac{\overline{X}_1 - \overline{X}_2}{s_p\sqrt{\frac{1}{n_1} + \frac{1}{n_2}}}$$

(since, in this case $\sigma$ *is unknown* and the two samples [men and women] are *independent*).

The data from Table 11.3 are used in the calculations of the following sample statistics:

$\overline{X}_1$ = the mean after treatment anxiety estimate for the men,

$\overline{X}_2$ = the mean after treatment anxiety estimate for the women,

$s_1$ = the standard deviation of the after treatment anxiety estimates for the men,

$s_2$ = the standard deviation of the after treatment anxiety estimates for the women, and

$s_p$ = the pooled standard deviation of the after treatment anxiety estimates for the men and women.

For the men:

| **Patient/ Gender** | **Anxiety Estimates After Treatment** $X_1$ | $X_1^2$ |
|---|---|---|
| 1/M | 06 | 36 |
| 2/M | 06 | 36 |
| 3/M | 06 | 36 |
| 4/M | 08 | 64 |
| | $\Sigma X_1 = 26$ | $\Sigma X_1^2 = 172$ |

$$\overline{X}_1 = \frac{\Sigma X_1}{n} = \frac{26}{4} = 6.5$$

$$s_1 = \sqrt{\frac{\Sigma X_1^2 - \left[\frac{(\Sigma X_1)^2}{n}\right]}{n-1}}$$

$$= \sqrt{\frac{172 - \left[\frac{(26)^2}{4}\right]}{3}}$$

$$= \sqrt{\frac{172 - 169}{3}}$$

$$= \sqrt{\frac{3}{3}}$$

$$s_1 = 1.0.$$

For the women:

| **Patient/ Gender** | **Anxiety Estimates After Treatment** $X_2$ | $X_2^2$ |
|---|---|---|
| 5/F | 08 | 64 |
| 6/F | 05 | 25 |
| 7/F | 08 | 64 |
| 8/F | 06 | 36 |
| | $\Sigma X_2 = 27$ | $\Sigma X_2^2 = 189$ |

$$\overline{X}_2 = \frac{\Sigma X_2}{n} = \frac{27}{4} = 6.75$$

$$s_2 = \sqrt{\frac{\Sigma X_2^2 - \left[\frac{(\Sigma X_2)^2}{n}\right]}{n - 1}}$$

$$= \sqrt{\frac{189 - \left[\frac{(27)^2}{4}\right]}{3}}$$

$$= \sqrt{\frac{189 - 182.25}{3}}$$

$$s_2 = \sqrt{\frac{6.75}{3}}$$

$$s_2 = 1.5.$$

For the men and women:

$$s_p = \sqrt{\frac{[s_1^2(n_1 - 1)] + [s_2^2(n_2 - 1)]}{n_1 + n_2 - 2}}$$

$$s_p = \sqrt{\frac{[1.0^2\,(4 - 1)] + [1.5^2\,(4 - 1)]}{4 + 4 - 2}}$$

$$s_p = \sqrt{\frac{3 + 6.75}{6}}$$

$s_p = 1.27$. Therefore,

$$t = \frac{6.5 - 6.75}{1.27\sqrt{\frac{1}{4} + \frac{1}{4}}}$$

$$= \frac{-.25}{1.27\sqrt{0.5}}$$

$$= \frac{-.25}{1.27(.7)}$$

$$t = -0.28$$

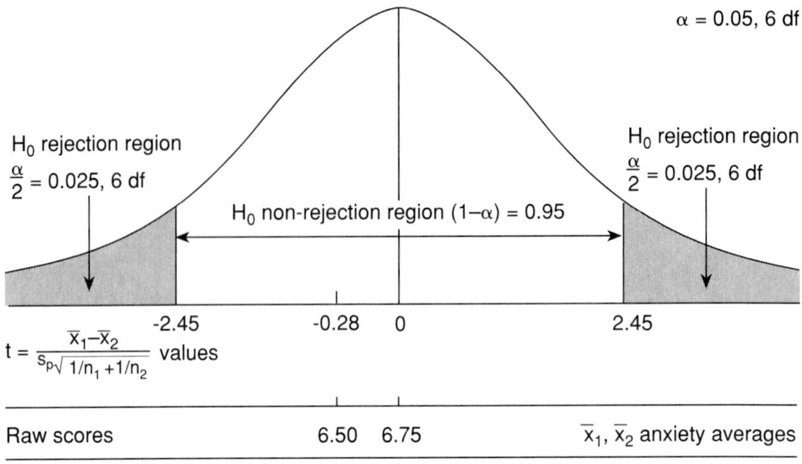


**FIGURE 11.5**
The critical t values (±2.45) associated with $\alpha = 0.05$ with 6 degrees of freedom and a nondirectional, two-sided test of the null hypothesis.

### *Step 3 Determining the null hypothesis rejection region:*

In Figure 11.5 we can see that the calculated t value $(-0.28)$ does not fall in the null hypothesis rejection region ($\alpha/2 = 0.025$, 6 df) for this two-sided test.

### *Step 4 Making a conclusion:*

Since the calculated t value does not fall in the null hypothesis rejection region, we fail to reject the null hypothesis. We conclude that the subjective anxiety estimates made by the male patients after treatment were *not* significantly different than the subjective anxiety estimates made by the female patients after treatment (at $\alpha = 0.05$, 6 df).

CHAPTER

12

# Linear Correlation: Describing Relationships

*"Neighborhood . . . is correlative, so that no man can be neighbor to another without the other being likewise neighbor to him."*

**TUCKER (1768)**

Correlation is the "co-relationship" between two variables. Karl Pearson (1857–1936), an English mathematician and scholar, produced the correlation coefficient that we know and refer to as the **Pearson r.** Pearson used his correlation coefficient to describe relationships between human characteristics, such as height and weight, in relative terms rather than in units of measurement, such as inches and pounds. Pearson studied the relative positions of the heights of fathers and the relative positions of the heights of their sons at maturity. The paired variables used by Pearson (the height of a father and the height of his son) were compared relative to how much each height varied from the average of its own group. That is, how did the father's height compare with the average height of the fathers, and how did the son's height compare with the average height of the sons?

Pearson's comparisons of relative position also made it possible to create a method of comparison between qualitatively different measurements. For example, measurements of height and weight cannot be compared directly because the two are conceptually different and they are expressed as different units. But they can be compared in terms of how much each varies from its own average. Therefore, the relationship between two qualitatively different variables can be expressed quantitatively. Pearson's correlation coefficient is, then, the numerical description of the relationship between two or more variables. A single value, the Pearson r, can be calculated to describe this relationship. In this chapter we will use the Pearson r to describe the relationship that exists between two variables (bivariate correlation analysis) only.

## THE INDICATION OF STRENGTH AND DIRECTION

The Pearson r, as with other correlation coefficients, describes the strength and the direction of the relationship between the variables. The strength of the relationship between the variables is described by a single number. This number may be equal to any value from +1.00 to −1.00, including zero.

A correlation coefficient never is greater than +1.00 or less than −1.00. A correlation coefficient of +1.00 or −1.00 indicates that the two variables are correlated to the maximum degree, or perfectly correlated.

A correlation coefficient of zero indicates that there is no correlation, whatsoever, between the two variables. When the correlation coefficient is equal to zero, the two variables are associated randomly.

The strength of the correlation is described by a single number, such that the closer the number is to +1.00 or −1.00, the stronger the degree of correlation between the variables. Therefore, a correlation coefficient of +0.85 is stronger than one of +0.65, and a correlation coefficient of −0.90 is stronger than one of −0.30. A correlation coefficient of +1.00 is equal in strength to one of −1.00. Likewise, correlation coefficients of +.82 and −.82 are equal in strength.

In social research, perfect correlations (±1.00) are unusual. In math and physics, perfect correlations occur more frequently. For example, a circle's diameter bears an exact (perfect) relationship with its circumference. Circles with larger diameters will always have larger circumferences, hence a perfect correlation coefficient of +1.00. In those disciplines that involve the observation of living things, perfect correlations are very unlikely.

The direction of the correlation is indicated by the sign of the correlation coefficient. The sign is not an indication of strength. That is to say, a negative correlation coefficient does not mean a lack of correlation; it simply indicates that the variables are correlated in an inverse fashion. When one variable increases, the other tends to decrease. While a strong *positive* correlation might exist between calories consumed and weight gained, a strong *negative* correlation might exist between muscle mass and age.

### The Scatter Diagram

Perhaps the easiest way to conceptualize correlation is to use the scatter diagram. The **scatter diagram** is a graph of the pairs of variables, using X and

**Scatter diagram 1**

**Data set 1**

| Pairs | Variables X | Y |
|---|---|---|
| 1. | 1 | 3 |
| 2. | 2 | 4 |
| 3. | 3 | 5 |
| 4. | 6 | 8 |
| 5. | 8 | 10 |

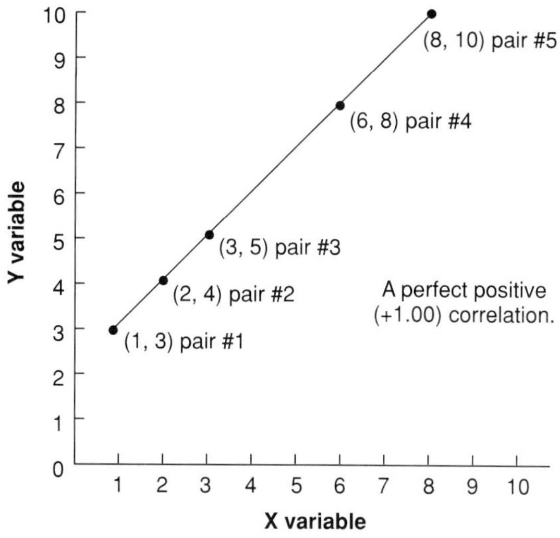


**FIGURE 12.1**
Data set 1 and the associated scatter diagram indicating the perfect positive linear correlation.

Y, to represent the two paired variables. For Pearson r, the paired variables are graphed using the interval or the ratio scale of measurement on both the X and the Y axis.

In Figure 12.1 the data and scatter diagram associated with a perfect *positive* linear correlation are presented.

For each pair, smaller values of X are associated with smaller values of Y and larger values of X are paired with larger values of Y. Therefore, for all the pairs, as X increases so does Y. Notice that a single straight line passes through each of the X, Y plots. These data are perfectly, positively correlated, with a correlation coefficient of $+1.00$.

In Figure 12.2 the data and scatter diagram associated with a perfect *negative* correlation are presented.

For each pair, smaller values of X are associated with larger Y values, and larger X values are paired with smaller Y values. Therefore, for all the pairs, as X increases, Y decreases. A single straight line passes through each of the X, Y plots. These data are perfectly, negatively correlated, with a correlation coefficient of $-1.00$.

In Figure 12.3 we see the data and scatter diagram associated with a weak linear, nearly random relationship between the plotted X, Y pairs. Notice that there seems to be no clear relationship, either positive or negative, among the plotted pairs. These data demonstrate a very weak (approaching zero), correlation.

**Scatter diagram 2**

**Data set 2**

| Pairs | Variables X | Variables Y |
|---|---|---|
| 1. | 1 | 10 |
| 2. | 2 | 8 |
| 3. | 3 | 6 |
| 4. | 4 | 4 |
| 5. | 5 | 2 |

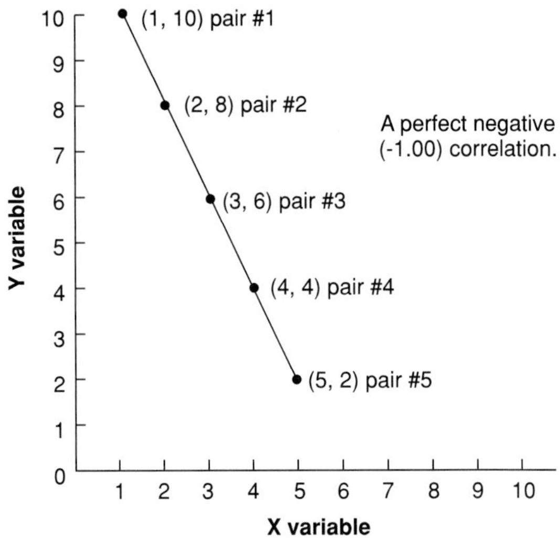


**FIGURE 12.2**
Data set 2 and the associated scatter diagram indicating the perfect negative correlation.

**Scatter diagram 3**

**Data set 3**

| Pairs | Variables X | Variables Y |
|---|---|---|
| 1. | 1 | 9 |
| 2. | 2 | 1 |
| 3. | 5 | 6 |
| 4. | 8 | 1 |
| 5. | 9 | 9 |

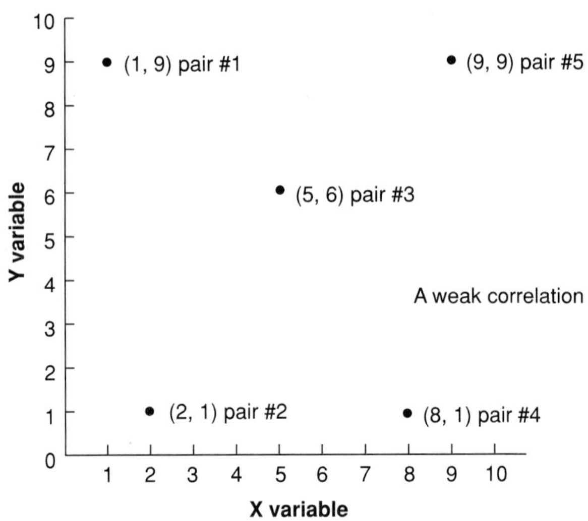


**FIGURE 12.3**
Data set 3 and the associated scatter diagram indicating a weak linear, nearly random relationship.

## THE CALCULATION OF THE PEARSON CORRELATION COEFFICIENT (r)

The Pearson correlation coefficient (r) is used when we want to describe the strength and direction of the relationship between two sets of paired numbers. The dependent variable (the outcome variable) is represented by X.

Usually, the pairs occur in one of two ways. First, the paired values often result from two different interval measurements on each of several cases (people or things). For example, we could record the body weight and the total serum cholesterol levels of several different cases. Our data set would be similar to the following set:

| **Cases (Person)** | **X (Body Weight)** | **Y (Total Serum Cholesterol)** |
|---|---|---|
| Al | 190 | 200 |
| Ed | 200 | 198 |
| Joe | 160 | 182 |
| Mike | 150 | 204 |
| Tom | 160 | 130 |

Second, one interval measurement is made on each member of several related cases. For example, we could measure the total serum cholesterol levels of several sets of fathers and sons. That data set would then be similar to the following set:

| **Pair (Father and Son)** | **X (Total Serum Cholesterol for the Father)** | **Y (Total Serum Cholesterol for the Son)** |
|---|---|---|
| 1 | 202 | 190 |
| 2 | 240 | 209 |
| 3 | 170 | 180 |
| 4 | 190 | 200 |
| 5 | 210 | 206 |

In this second case, the pairing is often based on some normally occurring, natural relationship, such as wives and husbands, sisters and brothers, identical twins, or roommates. Or it might be based on a contrived initial matching according to one specific variable such as resting heart rate, diastolic pressure, or weight.

A popular computational formula for the Pearson correlation coefficient (r) is

$$\text{Pearson r} = \frac{n[\Sigma XY] - [(\Sigma X)(\Sigma Y)]}{\sqrt{[n(\Sigma X^2) - (\Sigma X)^2]\,[n(\Sigma Y^2) - (\Sigma Y)^2]}}.$$

Where n = the number of pairs of observations:

$\Sigma XY$ = the sum of the products of the paired observations,

$\Sigma X$ = the sum of the observations of the independent variable, X,

$\Sigma Y$ = the sum of the observations of the dependent variable, Y,

$\Sigma X^2$ = the sum of the squared observations of the X variable,

$\Sigma Y^2$ = the sum of the squared observations of the Y variable.

Now, suppose a researcher wished to determine whether there was a relationship between the heights of fathers and the heights of their mature sons. The heights of 10 father-and-son pairs are recorded and the necessary sums are presented for use in the calculation of the Pearson r correlation coefficient. These data and calculations follow:

| **Father and Son Pairs** | **Height of Father X** | **$X^2$** | **Height of Son Y** | **$Y^2$** | **XY** |
|---|---|---|---|---|---|
| 1 | 72 | 5184 | 70 | 4900 | 5040 |
| 2 | 65 | 4225 | 69 | 4761 | 4485 |
| 3 | 69 | 4761 | 70 | 4900 | 4830 |
| 4 | 67 | 4489 | 69 | 4761 | 4623 |
| 5 | 71 | 5041 | 70 | 4900 | 4970 |
| 6 | 73 | 5329 | 72 | 5184 | 5256 |
| 7 | 70 | 4900 | 70 | 4900 | 4900 |
| 8 | 68 | 4624 | 70 | 4900 | 4760 |
| 9 | 64 | 4096 | 66 | 4356 | 4224 |
| 10 | 76 | 5776 | 74 | 5476 | 5624 |
| n = 10 | $\Sigma X = 695$ | $\Sigma X^2 = 48{,}425$ | $\Sigma Y = 700$ | $\Sigma Y^2 = 49{,}038$ | $\Sigma XY = 48{,}712$ |

Using these values and the computational formula,

$$\text{Pearson r} = \frac{n\Sigma XY - (\Sigma X)(\Sigma Y)}{\sqrt{[n\Sigma X^2 - (\Sigma X)^2][n\Sigma Y^2 - (\Sigma Y)^2]}},$$

we calculate

$$\text{Pearson r} = \frac{10 \times 48{,}712 - (695)(700)}{\sqrt{[10 \times 48{,}425 - (695)^2][10 \times 49{,}038 - (700)^2]}},$$

$$\text{Pearson r} = \frac{487{,}120 - 486{,}500}{\sqrt{[484{,}250 - 483{,}025][490{,}380 - 490{,}000]}},$$

$$= \frac{620}{\sqrt{[1225][380]}},$$

$$= \frac{620}{\sqrt{465{,}500}},$$

$$= \frac{620}{682.28},$$

$$\text{Pearson r} = +0.91.$$

**Scatter diagram 4**

**Fathers' heights (X), Sons' heights (Y)**

| Pairs | Variables X | Y |
|---|---|---|
| 1. | 72 | 70 |
| 2. | 65 | 69 |
| 3. | 69 | 70 |
| 4. | 67 | 69 |
| 5. | 71 | 70 |
| 6. | 73 | 72 |
| 7. | 70 | 70 |
| 8. | 68 | 70 |
| 9. | 64 | 66 |
| 10. | 76 | 74 |

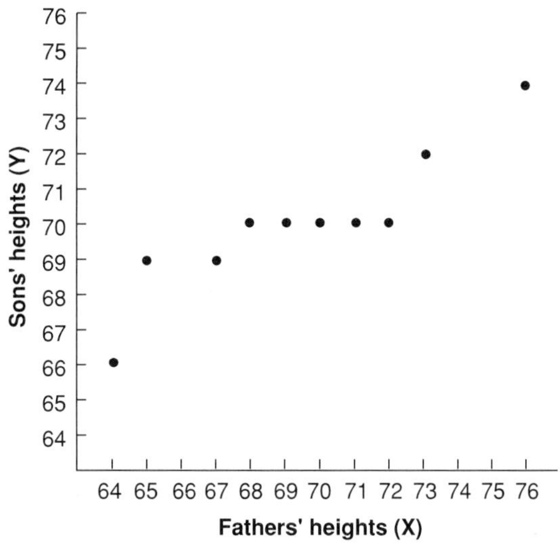


**FIGURE 12.4**
Data set 4 and the associated scatter diagram indicating a strong positive correlation between the heights of fathers and their sons.

Here the Pearson correlation coefficient, r = +0.91, describes the extent of linear correlation between the two continuous variables X and Y. The father's height (X) is the independent variable, and the son's height (Y) the dependent variable. The r value, +.91, is descriptive of a strong positive correlation between X and Y.

We can construct a scatter diagram of these paired data. In Figure 12.4 the data set and scatter diagram are displayed.

Notice that the X, Y plots, although not aligned perfectly, indicate a positive linear correlation between X and Y.

## Testing the Significance of the Pearson Correlation Coefficient (r)

One of two procedures may be used to test the significance of the calculated r value. Remember, we test the null hypothesis that the calculated r value = 0, *against* the two-sided alternative that the r value ≠ 0. When n (the number of pairs) is equal to or greater than 30, we can use a simple z test to determine the significance of r. When n is less than 30, we can use a t test to determine the significance of r.

Since our father and son example consisted of only 10 pairs, we must use a t test. The t statistic is

$$t = r\sqrt{(n - 2)/(1 - r^2)}.$$

If we use the calculated r value, +.91, from the father and son example, where n = 10, we have:

$$t = (+.91)\sqrt{\frac{10 - 2}{1 - (.91)^2}},$$
$$= (+.91)\sqrt{\frac{8}{1 - .83}},$$
$$= (+.91)\sqrt{\frac{8}{.17}},$$
$$= (+.91)\sqrt{47.06},$$
$$= (+.91)(6.86),$$
$$= 6.24.$$

For this particular test we have a calculated t value of 6.24. In this case, the degrees of freedom is two fewer than the number of pairs (10 − 2 = 8 df). We can now use the t table (Appendix Table B) to determine that a t value larger than ±2.306 is significant at the $\alpha = 0.05$ level when the alternative hypothesis suggests a two-sided test, 0.05/2 = 0.025. Remember, the $\alpha$ = 0.05 region is divided by 2 to form both $H_0$ rejection regions, one for each end of the distribution. Therefore, we actually have two $H_0$ rejection regions, each equal to 0.025 (0.05/2 = 0.025). We use the 0.025 $t_\alpha$ values at 8 df to find the critical t values, ±2.306.

Therefore, our calculated t value, 6.24, is larger than the critical t values ±2.306 at $\alpha = 0.05$ with 8 df, using the two-sided test of significance. Therefore, we reject the null hypothesis ($H_0$: r = 0.00) and conclude that the Pearson r = 0.91 indicates a significant correlation between the heights of fathers and the heights of their sons.

In the other situation, when n is 30 or greater, we can compute a z statistic to determine the significance of r. For this example, suppose our Pearson correlation coefficient (r) is equal to −0.70, and that 50 pairs were studied (n = 50). Then, to use z to test for the significance of r = −0.70 at $\alpha$ = 0.05, using a two-sided alternative hypothesis ($r \neq 0$), we have

$$z = (r)\sqrt{n - 1},$$
$$= (-0.70)\sqrt{50 - 1},$$
$$= (-0.70)\sqrt{49},$$
$$= (-0.70)(7),$$
$$z = -4.90.$$

From the z table (Appendix Table A) we can determine that a z value greater than ±1.96 indicates significance when $\alpha = 0.05$, using a two-sided test of the null hypothesis. Therefore, the Pearson r value of −0.70 is significant at $\alpha = 0.05$ using a two-sided test.

**Scatter diagram 5**

**Data set 5**

| Pairs | Variables X | Y |
|---|---|---|
| 1. | 1 | 1 |
| 2. | 2 | 3 |
| 3. | 3 | 5 |
| 4. | 4 | 7 |
| 5. | 5 | 9 |
| 6. | 6 | 8 |
| 7. | 7 | 6 |
| 8. | 8 | 4 |
| 9. | 9 | 2 |
| 10. | 10 | 1 |

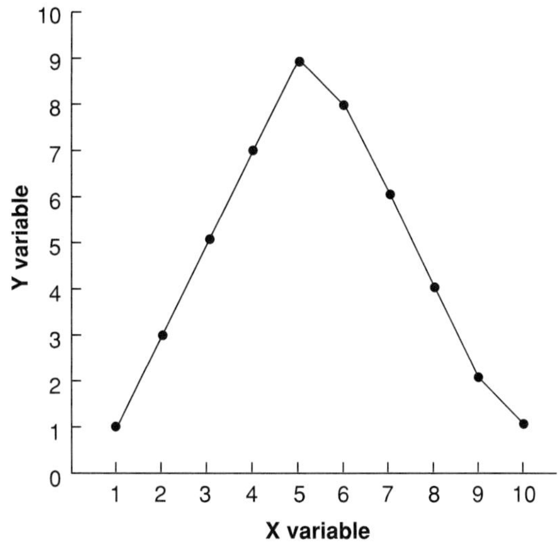


**FIGURE 12.5**
Data set 5 and the associated scatter diagram indicating a nonlinear relationship between X and Y.

## IMPORTANT INTERPRETATION PRECAUTIONS

The correlation coefficient is often misinterpreted. There are three commonly occurring misinterpretations of correlation.

First, correlation is used erroneously to indicate or infer cause and effect, that X causes Y. This is not necessarily true. When a significant correlation occurs, it only indicates that the two variables, X and Y, are associated or related. It is usually impossible to determine whether X influences Y, or whether either X or Y influences the other directly. In many cases it is a third variable that influences both X and Y, causing them to vary. For example, if a significant positive correlation exists between the independent variable (alcohol consumption), and the dependent variable (occurrence of throat cancer), it does not necessarily mean alcohol consumption causes throat cancer. It could be that a third variable, such as cigarette smoking—which is much more popular among those who consume alcohol frequently—is the cause.

Second, we erroneously use Pearson r to describe the degree of *linear* relationship between nonlinearly related pairs of data. One way to help determine the type of relationship present between the variables is to create a scatter diagram of the data.

The data set and scatter diagram describing a nonlinear relationship between X and Y are shown in Figure 12.5. In this case, if Pearson r was inappropriately calculated, it would result in a weak r value, thus indicating little linear correlationship between the X and Y variables. So, when considering the use of Pearson r with unfamiliar data, attempt to establish linearity, or the

**Scatter diagram 6**

**Data set 6**

| Pairs | Variables X | Y |
|---|---|---|
| 1. | 1 | 1 |
| 2. | 2 | 3 |
| 3. | 3 | 5 |
| 4. | 4 | 7 |
| 5. | 5 | 9 |

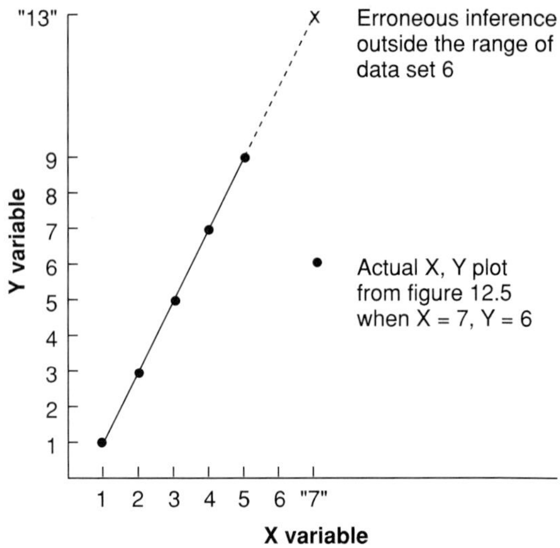


**FIGURE 12.6**
Data set 6 and the associated scatter diagram indicating a perfect positive linear relationship between X and Y.

lack of it, by creating a scatter diagram. Remember, a weak or insignificant r value only means the paired variables demonstrate little *linear* correlation. It may be that some other relationship exists. In the case of the relationship between the variables presented in Figure 12.5, a clear relationship does exist, it is just not a *linear* one.

Third, where we have a significant, positive, linear correlation, we are tempted erroneously to make interpretations *beyond* the range of the observations used in the calculation of Pearson r. To illustrate this error of interpretation, look at the five pairs of data presented in Figure 12.6. Notice that these pairs are the first five pairs of data from Figure 12.5. The scatter diagram suggests a perfect, positive correlation between the five pairs. The error occurs when we infer that this relationship (Figure 12.6) must also hold when X is 7 (outside the range of the five X values, 1 to 5). Therefore, we infer that when X = 7, Y must be equal to 13. In fact, when the rest of the data are included (Figure 12.5), we see that an actual X value of 7 is paired with a Y value of 6. It is an error to assume that the relationship that exists within a certain range of values continues beyond that range.

## Requirements For the Use of the Pearson Correlation Coefficient (r)

The use of the Pearson correlation coefficient is limited to paired variables that meet the following requirements:

1. The paired variables must consist of interval or ratio level data.
2. The relationship between the paired variables must be of a linear nature.

3. The sample providing the paired variables must be randomly selected from the population of interest.
4. The paired variables must not depart dramatically from normality. The shape of each variable's distribution must be somewhat symmetrical with but one mode.
5. The variation in the distributions of each of the paired variables must be about the same. If either of the paired variables is badly skewed, then Pearson r should not be used.

## FOR REVIEW

1. The Pearson correlation coefficient (r) is one of the most commonly used correlational statistical techniques. Pearson r is used to describe the degree of correlation or the strength of the correlation between two interval level, linearly related, paired variables. The degree or strength of the correlation is indicated by the r value, which may range from $-1.00$ to $+1.00$. The r value not only indicates the strength of the correlation, but also the direction of the relationship. A *negative* r value indicates an inverse relationship between the variables, while a *positive* r value indicates a positive relationship. A computational formula for the calculation of Pearson r is

$$\frac{n[\Sigma XY] - [(\Sigma X)(\Sigma Y)]}{\sqrt{[n(\Sigma X^2) - (\Sigma X)^2][n(\Sigma Y^2) - (\Sigma Y)^2]}}.$$

2. The scatter diagram is a graphic representation of plotted variable pairs, using X to represent the independent variable and Y to represent the dependent variable of each pair. The scatter diagram is used to determine the linearity (or lack of it) of the relationship between the plotted X, Y pairs.

### Chapter Knowledge and Comprehension Application

Before answering the following questions, consider each of the scatter diagrams in Figure 12.7.

1. Which of these scatter diagrams should probably not be described using Pearson r? Why?
2. Which of the scatter diagrams (is/are) *clearly* positive in direction?
3. Which scatter diagram seems to indicate the highest degree of correlation?
4. Which scatter diagram's r value would be closest to 0?

### Answers

1. Probably scatter diagram (d) and (e), because (d) seems to be a nonlinear relationship and (e) seems to indicate that the variation in variable X is dramatically greater than that of variable Y.

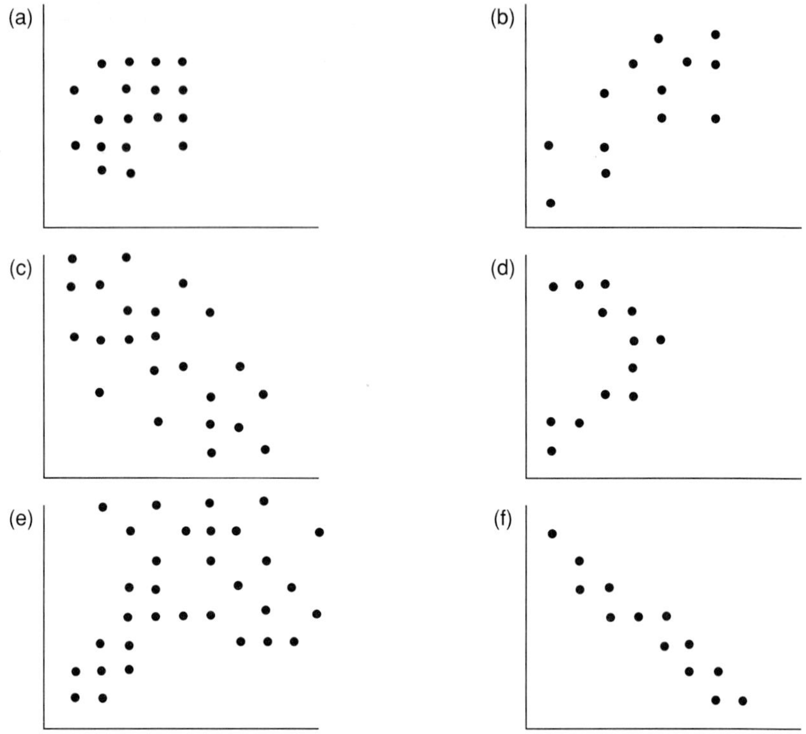


**FIGURE 12.7**
Six scatter diagrams (a–f). Assume the Y variable is represented by the vertical axis and the X variable is represented by the horizontal axis.

2. Clearly, scatter diagram (b) indicates a positive relationship.
3. Scatter diagram (f) seems to be the one that indicates the highest degree of correlation. A calculated r value for the pairs in (f) would probably be greater than −0.90.
4. Probably (a), because the plots seem to be nearly randomly located.

# A P P E N D I X

**TABLE A** Areas Under the Normal Curve

| z | .00 | .01 | .02 | .03 | .04 | .05 | .06 | .07 | .08 | .09 |
|---|---|---|---|---|---|---|---|---|---|---|
| 0.0 | .0000 | .0040 | .0080 | .0120 | .0160 | .0199 | .0239 | .0279 | .0319 | .0359 |
| 0.1 | .0398 | .0438 | .0478 | .0517 | .0557 | .0596 | .0636 | .0675 | .0714 | .0753 |
| 0.2 | .0793 | .0832 | .0871 | .0910 | .0948 | .0987 | .1026 | .1064 | .1103 | .1141 |
| 0.3 | .1179 | .1217 | .1255 | .1293 | .1331 | .1368 | .1406 | .1443 | .1480 | .1517 |
| 0.4 | .1554 | .1591 | .1628 | .1664 | .1700 | .1736 | .1772 | .1808 | .1844 | .1879 |
| 0.5 | .1915 | .1950 | .1985 | .2019 | .2054 | .2088 | .2123 | .2157 | .2190 | .2224 |
| 0.6 | .2257 | .2291 | .2324 | .2357 | .2389 | .2422 | .2454 | .2486 | .2517 | .2549 |
| 0.7 | .2580 | .2611 | .2642 | .2673 | .2704 | .2734 | .2764 | .2794 | .2823 | .2852 |
| 0.8 | .2881 | .2910 | .2939 | .2967 | .2995 | .3023 | .3051 | .3078 | .3106 | .3133 |
| 0.9 | .3159 | .3186 | .3212 | .3238 | .3264 | .3289 | .3315 | .3340 | .3365 | .3389 |
| 1.0 | .3413 | .3438 | .3461 | .3485 | .3508 | .3531 | .3554 | .3577 | .3599 | .3261 |
| 1.1 | .3643 | .3665 | .3686 | .3708 | .3729 | .3749 | .3770 | .3790 | .3810 | .3830 |
| 1.2 | .3849 | .3869 | .3888 | .3907 | .3925 | .3944 | .3962 | .3980 | .3997 | .4015 |
| 1.3 | .4032 | .4049 | .4066 | .4082 | .4099 | .4115 | .4131 | .4147 | .4162 | .4177 |
| 1.4 | .4192 | .4207 | .4222 | .4236 | .4251 | .4265 | .4279 | .4292 | .4306 | .4319 |
| 1.5 | .4332 | .4345 | .4350 | .4370 | .4382 | .4394 | .4406 | .4418 | .4429 | .4441 |
| 1.6 | .4452 | .4463 | .4474 | .4484 | .4495 | .4505 | .4515 | .4525 | .4535 | .4545 |
| 1.7 | .4554 | .4564 | .4573 | .4582 | .4591 | .4599 | .4608 | .4616 | .4625 | .4633 |
| 1.8 | .4641 | .4649 | .4656 | .4664 | .4671 | .4678 | .4686 | .4693 | .4699 | .4706 |
| 1.9 | .4713 | .4719 | .4726 | .4732 | .4738 | .4744 | .4750 | .4756 | .4761 | .4767 |
| 2.0 | .4772 | .4778 | .4783 | .4788 | .4793 | .4798 | .4803 | .4808 | .4812 | .4817 |
| 2.1 | .4821 | .4826 | .4830 | .4834 | .4838 | .4842 | .4846 | .4850 | .4854 | .4857 |
| 2.2 | .4861 | .4864 | .4868 | .4871 | .4875 | .4878 | .4881 | .4884 | .4887 | .4890 |
| 2.3 | .4893 | .4896 | .4898 | .4901 | .4904 | .4906 | .4909 | .4911 | .4913 | .4916 |
| 2.4 | .4918 | .4920 | .4922 | .4925 | .4927 | .4929 | .4931 | .4932 | .4934 | .4936 |
| 2.5 | .4938 | .4940 | .4941 | .4943 | .4945 | .4946 | .4948 | .4949 | .4951 | .4952 |
| 2.6 | .4953 | .4955 | .4956 | .4957 | .4959 | .4960 | .4961 | .4962 | .4963 | .4964 |
| 2.7 | .4965 | .4966 | .4967 | .4968 | .4969 | .4970 | .4971 | .4972 | .4973 | .4974 |
| 2.8 | .4974 | .4975 | .4976 | .4977 | .4977 | .4978 | .4979 | .4979 | .4980 | .4981 |
| 2.9 | .4981 | .4982 | .4982 | .4983 | .4984 | .4984 | .4985 | .4985 | .4986 | .4986 |
| 3.0 | .4987 | | | | | | | | | |

**TABLE B** Distribution of t, $t_\alpha$ Values for Various Degrees of Freedom

| | 1 | 2 | 3 | 4 | 5 | 6 | 7 | 8 | 9 | 10 | 11 | 12 | 13 | 14 | 15 | 16 | 17 | 18 | 19 | 20 | 21 | 22 | 23 | 24 | 25 | 26 | 27 | 28 | 29 | ∞ |
|---|---|---|---|---|---|---|---|---|---|---|---|---|---|---|---|---|---|---|---|---|---|---|---|---|---|---|---|---|---|---|
| 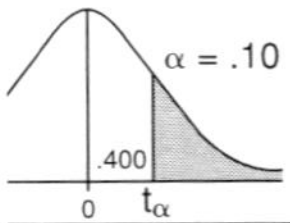 $\alpha$ = .10 | 3.078 | 1.886 | 1.638 | 1.533 | 1.476 | 1.440 | 1.415 | 1.397 | 1.383 | 1.372 | 1.363 | 1.356 | 1.350 | 1.345 | 1.341 | 1.337 | 1.333 | 1.330 | 1.328 | 1.325 | 1.323 | 1.321 | 1.319 | 1.318 | 1.316 | 1.315 | 1.314 | 1.313 | 1.311 | 1.282 |
| 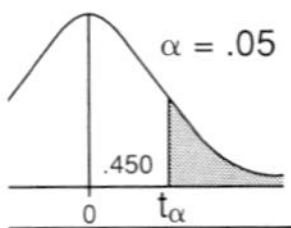 $\alpha$ = .05 | 6.314 | 2.920 | 2.353 | 2.132 | 2.015 | 1.943 | 1.895 | 1.860 | 1.833 | 1.812 | 1.796 | 1.782 | 1.771 | 1.761 | 1.753 | 1.746 | 1.740 | 1.734 | 1.729 | 1.725 | 1.721 | 1.717 | 1.714 | 1.711 | 1.708 | 1.706 | 1.703 | 1.701 | 1.699 | 1.645 |
| 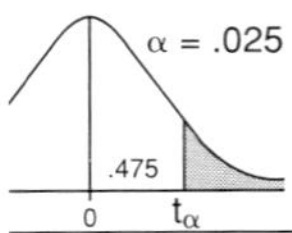 $\alpha$ = .025 | 12.706 | 4.303 | 3.182 | 2.776 | 2.571 | 2.447 | 2.365 | 2.306 | 2.262 | 2.228 | 2.201 | 2.179 | 2.160 | 2.145 | 2.131 | 2.120 | 2.110 | 2.101 | 2.093 | 2.086 | 2.080 | 2.074 | 2.069 | 2.064 | 2.060 | 2.056 | 2.052 | 2.048 | 2.045 | 1.960 |
| 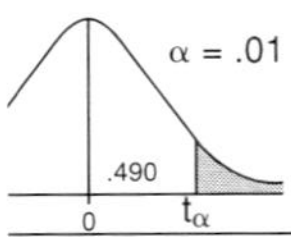 $\alpha$ = .01 | 31.821 | 6.965 | 4.541 | 3.747 | 3.365 | 3.143 | 2.998 | 2.896 | 2.821 | 2.764 | 2.718 | 2.681 | 2.650 | 2.624 | 2.602 | 2.583 | 2.567 | 2.552 | 2.539 | 2.528 | 2.518 | 2.508 | 2.500 | 2.492 | 2.485 | 2.479 | 2.473 | 2.467 | 2.462 | 2.326 |
| 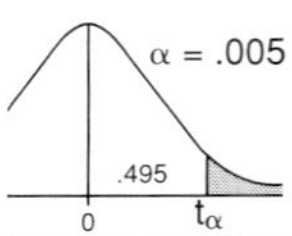 $\alpha$ = .005 | 63.657 | 9.925 | 5.841 | 4.604 | 4.032 | 3.707 | 3.499 | 3.355 | 3.250 | 3.169 | 3.106 | 3.055 | 3.012 | 2.977 | 2.947 | 2.921 | 2.898 | 2.878 | 2.861 | 2.845 | 2.831 | 2.819 | 2.807 | 2.797 | 2.787 | 2.779 | 2.771 | 2.763 | 2.756 | 2.576 |
| df | 1 | 2 | 3 | 4 | 5 | 6 | 7 | 8 | 9 | 10 | 11 | 12 | 13 | 14 | 15 | 16 | 17 | 18 | 19 | 20 | 21 | 22 | 23 | 24 | 25 | 26 | 27 | 28 | 29 | ∞ |

# INDEX